Seeds of Occupation, Seeds of Possibility

andrea noelani brower

seeds of occupation, seeds of possibility

The Agrochemical-GMO Industry in Hawaiʻi

West Virginia University Press
Morgantown

First edition published 2022 by West Virginia University Press
Printed in the United States of America

ISBN 978-1-952271-69-4 (paperback) / 978-1-952271-70-0 (ebook)

Library of Congress Cataloging-in-Publication Data
Names: Brower, Andrea Noelani, author.
Title: Seeds of occupation, seeds of possibility : the agrochemical-GMO industry in Hawai'i / Andrea Noelani Brower.
Description: First edition. | Morgantown : West Virginia University Press, 2022. | Series: Radical natures | Includes bibliographical references and index.
Identifiers: LCCN 2022017937 | ISBN 9781952271694 (paperback) | ISBN 9781952271700 (ebook)
Subjects: LCSH: Seed industry and trade—Hawaii. | Transgenic plants—Hawaii. | Agricultural chemicals industry—Hawaii. | BISAC: BUSINESS & ECONOMICS / Industries / Agribusiness | SCIENCE / Biotechnology
Classification: LCC HD9019.S432 U625 2022 | DDC 338.1/7—dc23/eng/20220414
LC record available at https://lccn.loc.gov/2022017937

Cover and book design by Than Saffel / WVU Press
Cover image: Hanohano Naehu at the Mana March on Kauai in 2013. Photo courtesy of the author.

Contents

Acknowledgments

This book is rooted in political movements, and my greatest debt is to my comrades. The wisdom, friendship, and fortitude of co-activists, too many to name, motivate and inform these pages.

My family is my anchor and joy. They are an inexhaustible reservoir of love and support. Koru and Teah, you are constant reminders of the boundless compassion and goodness of the human spirit.

This book is dedicated to West Kaua‘i and the innumerable communities globally who are impacted by the toxic products the agrochemical-seed-biotech industry pushes in their merciless quest for profits.

This book is derived in part from articles published in the journals *Capitalism Nature Socialism* (“Hawai‘i: ‘GMO Ground Zero,’ ” 2016) and *Food, Culture, and Society* (“From the Sugar Oligarchy to the Agrochemical Oligopoly: Situating Monsanto and Gang’s Occupation of Hawai‘i,” 2016), available online at https://doi.org/10.1080/10455752.2015.1112420 and https://doi.org/10.1080/15528014.2016.1208342, respectively.

Introduction

Contested Futures

In the fall of 2013, thousands of residents on the rural island of Kauaʻi marched on its County Council Chambers chanting "Aloha ʻāina,"* "Pass the bill," and "Stop poisoning us, the Garden Island's had enough." The moment likely marked the island's largest political demonstration in history (D'Angelo 2013). Under the banners of doctors, nurses, teachers, unions, mothers, Native Hawaiians, farmers, environmentalists, and surfers, the Mana March united diverse participation around demands for greater protection from pesticide use by the world's largest chemical corporations. Five weeks later, the Kauaʻi County Council passed unprecedented local legislation mandating pesticide disclosure, buffer zones, and a health and environmental study on the impacts of Dow, DuPont, BASF, and Syngenta's GMO (genetically modified organism) field testing and seed growing operations.†

Following Kauaʻi's historic regulation, Hawaiʻi Island County passed its own law that would prevent the agrochemical-seed-biotech companies from ever establishing operations on that island. Most remarkably, one year later Maui, Molokaʻi, and Lānaʻi islands passed a ballot initiative placing a temporary moratorium on "all GE practices and operations" pending an environmental and health impact study. Dubbed the "Maui miracle," the initiative passed with 50 percent of the vote despite Dow and Monsanto outspending advocates more than 100 to 1 with a state record of at least $8 million in direct advertising alone.

The bold local activism and laws that swept across the island chain in 2013 and 2014 caught the chemical industry by surprise. Accustomed to conflict in the game of regulatory evasion but not to losing, the industry mobilized rapidly

---

* Though most typically translated as "love for land," aloha ʻāina is a deeply spiritual and distinctly political concept with a long history in Hawaiian anti-colonial struggle (Silva 2004).

† The measure also applied to Hawaiʻi's largest coffee plantation, because it exceeded an annual use of five pounds or fifteen gallons of a single restricted pesticide.

to deepen and extend its local power and influence. Full-time propagandists were hired, government regulators were recruited for high-paying positions with the industry, astroturf groups exploded, progressive county politicians were replaced by industry employees through vicious slander campaigns, and the chemical companies sued to block all three ordinances in federal court. Despite industry's best efforts, in 2018 Hawai'i became the first state to ban the highly dangerous neurotoxin chlorpyrifos and adopted tepid pesticide disclosure and buffer zone measures. Due to public pressure, companies have also begun to reduce their operations near residential areas. The movement continues to build in its demands for greater protection from pesticide use, while also broadening its articulation of the systemic nature of social and environmental injustice. The occupation of Hawai'i by the agrochemical-seed-biotech industry has become one node of resistance in a growing movement that understands intersectionality of oppression and forges intersectionality of struggle. In this regard, it is deeply hopeful.

This book is a story of rising people's movements and the horizons of possibility they are opening, as well as countermovements that aim to blunt resistance and solidify the status quo. It is also the longer story of how the agrochemical industry came to occupy Hawai'i in the first place. With a history in plantation agriculture and oligarchic rule, Hawai'i is an epicenter of Bayer (Monsanto), DowDuPont (Corteva), ChemChina (Syngenta), and BASF's global chains of production. Dubbed GMO Ground Zero by activists, the most geographically isolated islands in the world have hosted more experimental field tests of genetically engineered (GE) crops than any other state in the US. These activities are almost entirely the domain of the four largest agrochemical-seed-biotech conglomerates, which function as a global oligopoly. Their operations in Hawai'i are pesticide intensive, use and pollute public lands and waters illegally taken from Kānaka Maoli (Native Hawaiians), and are premised upon ongoing US occupation of the islands.

The industry, backed by the popular press and the State of Hawaii, claims that it chooses to locate in the islands due solely to the "natural competitive advantage" of a warm and sunny climate. While not untrue—Hawai'i's year-round growing season is certainly favorable for speeding up the development of herbicide-resistant seeds and testing other agricultural technologies—this simplistic narrative obscures fundamental social and historical context. Further, the oft-repeated assertion that industry "contributions . . . are at no cost to the State" (Loudat and Kasturi 2013, 4) neglects mention of wide-ranging government supports and subsidies, and to whom the contributions and costs of such arrangements accrue. Such omissions are never politically neutral.

In contrast, this book offers a critical reading of why the most geographically isolated islands in the world have become central to a global agrochemical-seed-biotech oligopoly's chains of production. The conditions that make way for the industry's occupation of Hawai'i are not merely natural and, likewise, not inevitable. In other words, it is more than sunshine that makes Hawai'i's soils ideal for growing patented seeds engineered to withstand pesticides. Obfuscation of the underlying, socially determined conditions that give rise to the situation naturalizes plantations, oligarchies, inequality, and injustice. In a critical analysis of the reasons behind the agrochemical industry's establishment in Hawai'i, rather than sunshine, competitive market advantage, and other claims of innateness, what emerges are capitalist dispossession, imperial violence, local state facilitation, structural racism, and constant maneuvering by the powerful to maintain their interests. To demystify and denaturalize these socially created conditions is to assert the possibility of a break in this trajectory.

The agrochemical-seed-biotech industry emerges out of the past decades' intensification of capitalist logic, which has been a breeding ground for oligopolistic transnational corporations. The conditions for its existence are imperial-capitalist policies of the United States and expropriation of what previously belonged to us all, the commons. Its technologies are made through privatization of public science and serve the singular goal of maximizing corporate profits through entrenchment of a pesticide-intensive agricultural system. Through amassed wealth, these major capitalist conglomerates hold political power that in cyclical fashion engineers the ever-growing concentration of wealth. These are the subjects of the following two chapters.

The oligopoly's local occupation of Hawai'i is likewise conditioned upon a colonial relationship to the US state and capital, including a social and physical landscape descended from oligarchic sugar plantation days. Sugar's infrastructures, institutions, and ideas have been directly inherited by the agrochemical industry. Consolidated resource control and power—reflected in and reinforced by the local state—facilitate the industry's operations. Public lands (rightfully belonging to Native Hawaiians), waters, infrastructures, tax dollars, and research and educational institutions are diverted from other possibilities to support the industry. All of this takes place in the context of, and is entirely reliant upon, the ongoing occupation of Hawai'i by the United States following the illegal overthrow of a sovereign indigenous government.

The benefits and burdens of the industry's operations are grossly uneven and deeply racialized. Migrant workers, Native Hawaiians, and plantation communities shaped by sugar and pineapple face the highest exposure to dangerous pesticides used by the industry. As with other sites of environmental injustice, those living, working, and going to school near agrochemical fields experience a "miasma of uncertainty" (Steingraber 1997, 71). There is no mandatory disclosure of when, where, and what types of pesticides are used, making the extent of resident exposure unknown and complicating study of impacts. Elevated rates of pesticide-related health conditions and multiple incidents of school and worker poisonings demand information, study, and regulation that both the industry and the State of Hawaii have thwarted. The manufacturing of doubt has been a primary tactic to delay regulatory action and avoid culpability. Chapter 3 provides an overview of the landscape of what activists dub "GMO Ground Zero," while chapters 4 and 5 detail the history that brought the islands to this moment. Chapter 6 continues with the question of why the industry chooses to locate in Hawai'i, with particular focus on the role of government and the extensive public supports received by private capital.

Many in Hawai'i believe in and are mobilizing for alternative futures. Bold regulations have been passed in three of four island counties despite record industry spending against them. Though those county laws were blocked by industry lawsuits, the movement continues to make remarkable strides. It has changed public understanding, galvanized widespread political participation, forced passage of precedent-setting state pesticide laws, and built wider solidarity across issues. In contrast to consumer-focused GMO concerns in much of the US, Hawai'i's movement draws attention to producing communities, the inextricability of pesticides and current uses of genetic engineering, and the need for collective political engagement to combat structural injustice. Chapter 7 gives an account of these events and their significance locally and globally.

Predictably, the rise of mass people's movements in Hawai'i has become a major concern to the industry. Chapter 8 examines the regressive countermovement, in both its specific and its wider formations. From within a moment marked by the dominance of neoliberal norms, agrochemical operations are made to appear inevitable—as the lesser of evils in a predestined monoeconomy of haves and have-nots. To be "realistic" in the severely narrowed landscape of neoliberal possibility is to seek "consensus" with chemical monopolies, to consent to their voluntarism, and to allow capital and its political support to define the conditions of our collective lives. The engagement of people's movements to challenge and change such conditions—the policies and systems that define our lives—is deemed messy, conflictual, unruly, objectionable, and ultimately futile.

For these reasons, the battle over what is considered possible in the social order is a fundamental terrain upon which we struggle. Deep, transformative, systemic change must be up for consideration. When there is an ideological void of systemic alternatives to the status quo, it indeed does seem that agrochemical oligopolies and their environmental injustices are inevitable. Jobs and livelihoods appear to be opposed to human health and the environment. Real material limitations from within conditions of the present come to represent the entirety of social possibility. While movements typically seek to exchange local for global, or small for large, there is less attention to the social relations and structural limitations of the imperial-capitalist political economy. Much of this stems from a deep collective cynicism about the extent and depth of social change that is actually possible.

Hawai'i's agrochemical industry resistance movement contains hopeful seeds of transformative possibility. Imperialism and capitalism were rooted in the islands largely in relation to plantation agriculture. As agrochemical-GMO plantations are challenged, so are some of the foundations of the contemporary colonial-plantation economy. The movement is central to a wave of organizing that is increasingly bringing together decolonial, environmental, labor, and economic justice struggle. There is great potential for development of more radical consciousness and organizing through "intersectionality of struggle" (Davis 2016). Values antithetical to capitalism and expression of alternative possibilities are continuing to grow. These are the topics of a final chapter on seizing the possible.

It is worth briefly stating for readers what this book is not. While written from within struggles around the agrochemical industry in Hawai'i, it is not a thorough overview or interrogation of that diverse resistance. It is also not a complete history of plantations in Hawai'i. Though plantations and resistance are central topics, this book does not aim to tell a comprehensive story of either. Further, while this book centers on what activists describe as "GMO Ground Zero," it is not about agricultural biotechnology per se. It does not detail all of the concerns or debates around the technology or draw broad conclusions about it. In discussing agricultural biotechnology, what is centered are questions of capitalism, imperialism, power, ownership, and control, and what these things mean for democracy, ecology, equality, and justice. In regard to terminology, "GE" is used as an abbreviation for "genetically engineered." However, rather than "GEO" (genetically engineered organism), "GMO" (genetically modified organism) is used because it is the common vocabulary in Hawai'i. While even the definition of agricultural biotechnology is intentionally confused and contested by industry, it is used in this book consistent with the predominant meaning of manipulating an organism's genetic material through insertion of new pieces of DNA or by modification of base unit letters of the genetic code.

The island of Kaua'i figures centrally in the stories and analysis in this book. The primary reason for this is the author's own embeddedness in social movement struggle on that island. Broad-based resistance to agrochemical industry occupation erupted especially on the island of Kaua'i in 2013 and largely influenced what continued to build throughout the islands in the following years. Much of the research for this book was conducted during that year through activist ethnographic methods. The bias for Kaua'i is not intended to erase the struggle on other islands; it is merely a matter of the author's own limited knowledge.

Activist ethnography, the primary methodology used for this research, refers to immersion in social struggle as a participant, researcher, and theorist. It is a commitment to using knowledge and research to inform, expand, enliven, and convoke political struggle and imagination (Haiven and Khasnabish 2014). Rather than merely to describe, the aim is to effect. Such orientation runs counter to the purported apolitical stance of much academic research today. The stance of neutrality is a declaration of consent to the way things are, an alibi for inaction, and a fantasy of absolving oneself from the ways of the world. To take a lesson from the great Howard Zinn, "events are already moving in certain deadly directions, and to be neutral means to accept that" (2002, 8), or from the late anthropologist Eleanor Leacock, whose anti-racist and feminist work was consistent with the approach of activist ethnography, "To attempt neutrality . . . means to align oneself, by default, with the institutional structures that discriminate against and exploit" (Leacock 1987, 323). While many researchers no doubt intend neutrality, there is an inevitable stance taken in a researcher's selections and emphases (Zinn 1980). Against positivist assumptions about research, activist ethnography makes "no pretence of objectivity" (Graeber 2009, 33). For the activist scholar, it is both impossible and undesirable to be neutral in the conflicts of the world (Bisaillon 2012; Zinn 1990).

We are living in a moment of unparalleled inequality, of extreme human deprivation alongside lavish wealth. Climate breakdown and the sixth extinction are upon us. It does not need to be this way. Dismantling the naturalization of systemic cruelty and catastrophic environmental destruction is an imperative task of emancipatory politics and research. Thus, one of the core themes that motivates this book is how we seem stuck in our bleak present, confined in a logic of impossibility, of making only small changes here and there. While capitalist elites proclaim expansive possibility in techno-utopian futures, horizons of possibility in regard to the social order appear constricted and immutable. A social order of brutal hierarchy, inequality, and human suffering is continually reproduced in ways that give the appearance of inevitability, of there being no alternative, despite abundant evidence of our collective human capacities for something better. We create and recreate the world every day, but it has been coded unrealistic to believe we might actually create a world structured by logics of egalitarianism, cooperation, democracy, and ecological regeneration.

On the horizon are both deeply troubling and hopeful signs. Considerations of what is possible in the social order are expanding, sometimes very rapidly. Capitalism is up for contestation again in the years following the global financial crisis, Occupy, and uprisings around the world. US youth at the core of capitalist imperialism increasingly favor an overhaul of the political-economic system. The largest, most interracial protests in history are being recorded. A rising tide of social movements is articulating the intersectionality of struggle and demanding systemic change. Political candidates rooted in these movements are being brought to power by them. Self-described democratic socialist candidate for US president Bernie Sanders galvanized astonishing popular support and 80–90 percent of the under-thirty vote, despite desperate attempts to erase the campaign's political revolution message and militant edge. As it tends to, right-wing reactionism is being ignited parallel to leftist uprising. White supremacy, xenophobia, fear, exclusion, and authoritarianism are also galvanizing the masses. If one thing is clear, it is that much is up for grabs and it is incumbent upon us to think big.

Chapter 1

The Agrochemical-Seed-Biotech Oligopoly

The first task of this book is to make sense of the existence of a chemical+seed industry that is monopolistically controlled by a handful of corporations with a long history in war, poison, deceit and destruction. Four massive conglomerates control more than 60 percent of the global markets for commercial seeds and agricultural chemicals. They also control the market for virtually all genetically engineered crops. Far from an aberration, the oligopoly is the logical outcome of the structural dynamics of capitalism and especially neoliberal, hypercapitalist policy. The policies of neoliberalism, contra being a springboard for efficiency, innovation, competition, freedom, and less government, have instead been the breeding ground for monopolistic corporations that function transnationally by way of state support. The enormous profits being amassed by the agrochemical-seed-biotech industry are entirely conditioned upon theft of things previously considered common.

Capitalism and the Commons, in Brief

The logic of capital—competitive maximization of private profit—compels it to incessantly create ever new commodities, things that can be bought and sold on the market. Capitalism turns nature, products of human creativity and labor, humans, and other living beings themselves into commodities. The generation of new commodities often takes place through the enclosure or expropriation of what was previously deemed common, or belonging to many. Expropriation is a systemic compulsion, driven by capitalist market imperatives of competitive accumulation and profit maximization. A state can either put limits on the logic of capital or actively extend it.

Neoliberalism, the reigning ideology and political program of the past decades, is at its core about extending the logic of capital to as much of life as possible. Key to the neoliberal program has been a radical deepening of expropriation and enclosure of the commons. Modern enclosures extend the operational space of capitalism by commodifying new realms and creating market dependencies on things that were previously outside of the market even in predominantly capitalist economies (Sassen 2010, 25). This deepening of the reach of capital has required new conceptualizations of the nonhuman world and accompanying revolutions in law and policy (Heynen and Robbins 2005, 5). Through extreme legal and political changes, capitalist enclosures have encroached life-essential and previously noncommodifiable commons, extending the logic of the market into everything from pollination to the atmosphere, and drilling down to the basic building blocks of life.

Enclosure of "natural" commons provided freely by the earth—water, seeds, biodiversity, ecosystem services—are intertwined with expropriation of human-generated commons. Some have argued that such cultural or social commons are the main source of wealth production today (Virno 2004; Hardt and Negri 2009). Ideas, information, knowledge, relationships, technology, affects, and other fruits of human commoning are increasingly relied upon by capital for private wealth accumulation (Hardt 2011). Capitalist expropriation of cultural commons requires artificial exclusion where free access and dissemination might have been possible and available (Gorz 2010). For example, intellectual property right (IPR) mechanisms such as patents, copyrights, and trade secrets establish artificial scarcity where relations of open sharing might instead exist. It takes a political apparatus to manufacture and enforce scarcity and exclusivity, and prices of intellectual property reflect politics and power more than market conditions (Adair 2010). In other words, privatizing what was and could be common is entirely about the operation of power and serving the interests of some over others. While purportedly about less government, neoliberalism relies entirely on states to execute these processes of expropriation, enclosure, commodification, and marketization.

Enclosure of Seed Commons

Until very recently in human's agricultural history, people saved and shared seed, creating and recreating the crop and agroecosystem diversity that was considered common heritage and a foundation of social life (Mascarenhas and Busch 2006). In noncapitalist societies, decisions over seed saving and allocation were made within the overarching social arrangements and norms of which they were a part. While some customary arrangements recognized a degree of exclusivity in access to genetic resources, they largely operated on principles of reciprocity, gift exchange, and open sharing. Rather than restricting access, they aimed to propagate and spread seed and crop diversity (Kloppenburg 2010, 371).

Over the past century, forces of capitalism have expropriated seed commons for select private gain through legal and technological strategies. These trends began especially in the 1930s with the development of hybrid varieties of corn that were unfavorable for replanting (Kloppenburg 2004). To produce hybrid seeds, two distinct parental lines are needed. Thus, they can only be reproduced by the breeder. The technological ability to guarantee that farmers would have to repurchase seeds annually attracted enormous interest from private industry. By 1944, hybrid seeds generated revenues of $60–70 million (Paul and Steinbrecher 2003). As seed companies' power and influence over the political process grew parallel to their wealth, laws regarding plant breeders' rights were designed to grant them exclusive claims over crop varieties in which hybridization was not possible (Kloppenburg 2004). The US Congress was the first in the world to establish plant related intellectual property rights in 1930 with the Plant Patent Act, which was ratcheted up with the 1970 Plant Variety Protection Act to cover seeds themselves. The first international agreement on plant breeders' rights took place in 1961 with the International Convention for the Protection of New Varieties of Plants.

While extending private property protection into the domain of plants and seeds, the scope of these first IPR regimes was still limited. None went as far as preventing a farmer from saving seeds from past crops or restricting plant breeders from openly using patented seed varieties in their research (Matson, Tang, and Wynn 2012). Up until 1980 it was assumed that full utility patents—those issued for the invention of a new and useful process, machine, manufacture, or composition of matter—could not be granted for life forms. This changed with the landmark United States Supreme Court decision *Diamond v. Chakrabarty* (1980). The case began with General Electric seeking a patent on a bacteria that could digest oil, developed by one of its microbiologists, Ananda Chakrabarty. By his own testimony, Chakrabarty had "simply shuffled genes" (Breu 1980). A small number of public interest groups followed the case, opposing the patent on the grounds that "to justify patenting living organisms, those who seek such patents must argue that life has no 'vital' or sacred property . . . and that once this is accomplished, all living material will be reduced to arrangements of chemicals, or 'mere compositions of matter' " (Kimbrell 1996, 2). The court dismissed these concerns, ruling that "the issue was not whether there was a 'relevant distinction (in patentability) between living and inanimate things,' but whether living products could be seen as 'human-made inventions' " (Kimbrell 1996, 2).

Since that precedent-setting decision, patents on life forms have become a lucrative domain of capitalist enclosure. Five years after *Diamond v. Chakrabarty*, the US Board of Patent Appeals and Interferences ruled that the case could be extended to apply to utility patents on both seed varieties (germplasm) as well as transgenic traits (following the patenting of a mammal). Accordingly, by patenting a transgenic trait inserted into a crop, the entire plant is covered by strict property law. The US Supreme Court later held that such patents prohibit seed saving and research exemptions. Moreover, the scope of what is considered patentable has expanded to include research tools (versus only final products) and "discoveries" of already existing objects such as genes. In the years following *Chakrabarty* the line between what is considered "natural" and what is considered "invented" has shifted to a remarkable degree. As Harvard professor of science and technology studies Sheila Jasanoff writes, "only the most elemental physical phenomena such as electricity are today reliably regarded as not patentable" (2005, 209).

Plant breeders' rights cover plant varieties, while patents can additionally be applied to genes, gene sequences, breeding processes, and biological information. In the US, both are held for twenty years (Matson, Tang, and Wynn 2012). Full patents have also been issued for conventionally bred (not genetically engineered) plants and animals (Then and Tippe 2012). Despite the European Patent Office announcing that it would stop granting patents on conventional breeding in 2016, they have already granted around two hundred such patents (Global Agriculture 2017). Monsanto, for example, was issued a patent on conventionally bred melons originally from India, allowing them to block access to breeding material derived from the melon. However, genetically engineered crops are still associated with far higher levels of patent restrictions than conventionally bred crops because they enable an interlocking of multiple forms of intellectual property rights. A single genetically engineered plant can be covered by plant breeders' rights legislation and a utility patent simultaneously, while also being covered by utility patents on transgenic traits that have been inserted into the plant by means of genetic engineering. Further, inserted traits can be the property of multiple companies. In some countries where patents are not granted for plant varieties, the seed-biotechnology industry has been issued patents on traits, giving them roundabout patent rights over genetically engineered plant varieties. Methods of genetic engineering are also patentable. One study identified more than seventy patents embedded in a single variety of GMO golden rice (Kryder, Kowalski, and Krattiger 2000).

The patenting of seeds and genes is a process of converting naturally existing commons of the earth into private property. At the same time, it is expropriation and enclosure of collective human knowledge, cultural commons. Capitalist firms appropriate farmer and scientific knowledge to privatize and commodify the fruits of long histories of thinking and producing in common. Both conventional breeding and agricultural biotechnology depend entirely on quality gene lines developed within social institutions of cooperation and open sharing (Zerbe 2015). Capitalist firms' access to the prodigious pool of genetic diversity that has been produced over many millennia is facilitated by the systematic appropriation of plant varieties from farming communities around the world, their storage in gene banks and subsequent use in breeding programs. Genetic expropriations—almost entirely from poorer, indigenous, and communities of color—are often referred to as "biopiracy" (Mgbeoji 2014). Michael Hardt goes further in his criticism: "Piracy is actually a misnomer for such activities. Pirates have a much more noble vocation: they steal property. These corporations instead steal the common and transform it into property" (2011).

The exploitation of collective, multigenerational intellect also takes place in the realm of more formal science. Where biotechnology is involved, innovations are highly "polycentric and spasmodic," relying on many and involving past cooperatively generated knowledge (Jasanoff 2005, 207). In contrast, patents presume solitary and identifiable inventors and moments of invention in order to serve the function of capital accumulation. With biotechnology, most of these single moments of purported invention are actually premised upon cumulative publicly funded science, not innovations that have taken place primarily within the private sector. Even within the private sector, corporations themselves are highly collective entities, structured to harness the creative cognitive abilities of collaboration. Internally, biotechnology companies organize around common goals of advancing knowledge and technology. Knowledge is openly shared to achieve collective ingenuity and innovation, while the results of this commoning are then put to the service of very narrow private gain (see Charles 2001).

Achieving the extreme policy changes necessary to capitalist enclosure of seed and genetic commons has required imperial pressure of the US government, operating most notably in the realm of intellectual property right regimes. Strong IPR rules protect large US corporate advantages in a range of technology exports, including biotechnology, pharmaceuticals, and industrial and military electronics. They also raise barriers to entry against new and smaller enterprises, guaranteeing the dominance of already established monopolies (McAfee 2003). Rather than benefiting American people or the economy as a whole, the US government's pushing of strong IPR rules preferentially treats and actively redistributes wealth toward already dominant firms. Moreover, strong IPRs generally disadvantage poorer countries that lack the resources for research, development, and operation in the expensive and legally complex world of patents (Paul and Steinbrecher 2003). While the US claims that such policy advances technological innovation, US and other dominant firms vigorously oppose biotechnology transfer and benefit sharing with countries of the Global South (McAfee 2003).

Compulsions to adopt strict national laws in regard to intellectual property initially took place through the World Trade Organization's (WTO) Agreement on Trade-Related Intellectual Property Rights (TRIPS). During President Ronald Reagan's administration, a shift was made away from the United Nations' World Intellectual Property Organization (WIPO) and instead to the WTO. WIPO's format of one nation state, one vote—and the resulting attention to the concerns of poorer countries—was not conducive to the interests of American capital. In contrast, in the trade arena the US and American corporations were able to exert a high level of control over intellectual property governance. In *Private Power, Public Law*, Susan Sell documents how twelve powerful corporations that most stood to benefit from strict IPR regimes, including Monsanto, DuPont, Pfizer, and General Motors, succeeded in "getting most of what they wanted from a global IP agreement" in the drafting of TRIPS (2003, 2). TRIPS "ushered in a full-blown, enforceable global intellectual property regime that reaches deep into the domestic regulatory environment of states" (2003, 1). While both sweeping and legally binding, Sell argues that, in retrospect, TRIPS "looks like a relatively timid and permissive agreement" (2011, 448). Despite getting nearly everything they wanted in TRIPS, dominant corporate interests have pushed ever harder for extending their gains further. Over the past twenty years, bilateral and regional trade agreements have institutionalized even stricter and broader IPR standards.

Enforcement of intellectual property rights related to seeds and genes is strongest in countries like the United States, Canada, and Australia. Additionally, court rulings in these countries have indicated a tendency toward expansive interpretation of property rights in favor of the seed industry (Parfitt 2013). In the well-known case of canola breeder Percy Schmeiser, the Canadian Supreme Court ruled that Schmeiser had breached Monsanto's intellectual property rights when that company's patented genetically engineered canola was found in his fields. Schmeiser alleged that Monsanto's varieties had contaminated his fields and were not intentionally planted, but the court ruled that this was irrelevant. Similarly, US patent law does not require a showing of infringement intent. Often missed in the story of Schmeiser is that he was forced to abandon his own seed, which he and his wife had been saving for fifty years (Paul and Steinbrecher 2003).

In jurisdictions such as the United States, Australia, and Canada, farmers must sign technology use agreements with seed companies when purchasing genetically engineered seeds. These contracts impose patent license fees, as well as requirements to apply the company's brand of herbicide, to "comply with all reasonable directions and instructions" given by the seed company, to deliver crops to "approved grain handlers," and to not save seeds or supply seeds to others (Parfitt 2013, 49). Further, contracts require farmers to give companies access to their farm records and premises for a stated period after planting. In 2003 Monsanto had a staff of seventy-five dedicated to the task of catching farmers in breach of contract rules. DuPont had similarly hired private investigation firms to pursue farmers. As of 2013, Monsanto had filed 144 lawsuits based on claimed violation of technology use agreements in the US, involving 410 farmers and fifty-six farm companies. The number of investigations is far higher, as most cases are confidentially settled out of court. The total amount of money that Monsanto has collected from farmers through lawsuits and settlements is unknown, but exceeds $100 million just in what has been recorded on the company's own website (CFS and SOS 2013, 30).

Consolidation of the Agrochemical-Seed-Biotech Industry

Gene and seed enclosures go hand in hand with the rise of a global seed oligopoly that also controls the agricultural chemical and biotechnology markets. As of 2020 just four companies—Bayer (which acquired Monsanto in 2018), Corteva (the rebranding of Dow and DuPont's agricultural units), ChemChina (which acquired Syngenta in 2017), and BASF (which bought Bayer's seed division to satisfy antitrust regulators)—control over 60 percent of the global seed and 75 percent of the global agrochemical markets. The extent of concentration in what has become the global agrochemical-seed-biotech industry has reached an unprecedented high, though consolidation in the agricultural input industry is not a new phenomenon.

Prior to hybrid seed technologies, most commercial seed suppliers were small businesses that sold varieties developed in the public domain with a regional focus (Hubbard 2009). The US government sponsored programs to collect, develop, and distribute seeds for free, which was widely considered a public good that was vital to a stable and productive agricultural system (Matson, Tang, and Wynn 2012). Following the emergence of the private seed industry, the US public seed distribution program was abolished in 1924. As lucrative opportunities developed in hybrid seeds, consolidation of the industry advanced (Kloppenburg 2004). After World War II, private industry focused especially on developing hybrids for industrialized production, including varieties responsive to fertilizer and suitable for mechanical harvesting. Industrialized methods of dense monocrop production created ideal conditions for insects, disease, and weeds, thus requiring the use of agrochemicals. Postwar, increasing yields, as well as overproduction due to the gutting of supply management policies, led to downstream "innovations" of livestock feed, high-fructose corn syrup, other highly processed food fillers, and biofuels. All supplied further profit and consolidation opportunities for not only seed, agrochemical, fertilizer, and tractor companies but also for food processors, factory farms, and other big agribusiness.

By the 1960s, several large chemical and oil companies had acquired major seed companies (Srinivasan 2003). As the industry amassed wealth and political influence, more extensive patents for plant breeders were secured in the 1970s, spurring additional expansion of the seed industry. Companies began to integrate breeding, production, and conditioning and replace the public sector as a source of seed and knowledge (Fernandez-Conejo and Schimmelpfennig 2004). The advent of genetic engineering and the granting of full utility patents on living organisms following *Diamond v. Chakrabarty* initiated a new level of capital accumulation and consolidation.

Monsanto is most notable as an early big player. It largely pioneered strategies followed by other major chemical-pharmaceutical corporations. Monsanto was founded as a drug company in 1901. Its first product was saccharin for Coca-Cola, a derivative of coal tar that was later linked to bladder cancer. Prior to the creation of its agricultural division in 1960, Monsanto primarily manufactured chemicals, rubbers, and plastics (Hauter 2012). It later moved into agrochemicals, including Roundup and 2,4-D (an active ingredient in Agent Orange). In the early 1980s, with new opportunities to capitalize on both novel intellectual property rights and genetic engineering, Monsanto began to focus more on agricultural biotechnology (Charles 2001). In addition to changes in patent law, several factors played a role in the decision to increasingly seek profits in biotechnology. The termination of Monsanto's patent on its blockbuster herbicide Roundup was looming. At the same time, there was a perception that the chemical industry was a sunset industry vulnerable to the rising price of oil and competition from generics. New products were difficult and expensive to develop. Monsanto was also being forced by the growing environmental movement to abide by new regulations. Company executives feared that lawsuits over environmental contamination could go as far as bankrupting the company. Finally, the 1980 Bayh-Dole Act opened publicly funded research to private sector exploitation, which Monsanto took careful notice of (Glover 2010; Charles 2001).

Monsanto personnel became frequent visitors to university biotechnology labs. Upon seeing opportunity, they began their own research program, which was among the first to successfully genetically modify a plant cell. Research costs were high and required revenues from Roundup and other chemical products, triggering internal tensions in allotment of resources and decisions about the company's future direction (Charles 2001). It was the discovery of the ability to engineer seeds to be resistant to Monsanto's top-selling herbicide Roundup that solidified an apparent turn away from chemistry and toward biotechnology. In bringing Roundup-resistant crops to the market, Monsanto preserved the dominance of its most lucrative product by coupling proprietary Roundup Ready seeds with their brand of herbicide. The company promoted its turn to biotechnology as a radical break from the past, but from the start its genetic engineering ventures rested entirely on a combined agrochemical-GMO strategy (Glover 2010). Meanwhile, Monsanto spun off its industrial chemical division, which faced tremendous legal liabilities, and declared bankruptcy in 2003 (Hauter 2012).

For a short time, Monsanto developed genetic traits and licensed these to other companies to do the actual breeding and selling of seeds. Moving to further consolidate, in the 1990s a new strategy developed, and major chemical and pharmaceutical companies involved in biotechnology began to acquire seed businesses themselves (Srinivasan 2003). In acquiring seed companies, these already dominant corporations also obtained vital assets of well-developed seed lines and germplasm, production capacity, marketing and distribution links, and recognizable branding (Matson, Tang, and Wynn 2012). Over the next decade, Monsanto spent more than $12 billion buying at least thirty seed companies and several agricultural biotechnology firms (Khan 2013). In what Wenonah Hauter calls the "merger mania" of the 1990s, giant companies that branded themselves as being involved in "life sciences" emerged, all as major players in seeds, agrochemicals, and biotechnology (Hauter 2012, 236). Every large national seed firm in North America was acquired by one of these companies (Graff, Rausser, and Small 2003). Many retained their old company names, making the outstanding consolidation in the industry less visible to farmers and the public. At least two hundred independent seed companies were dissolved (Hubbard 2009).

The 1990s consolidations were driven to a remarkable extent by control over intellectual property rights. Vertical integration—controlling a range of capacities and intellectual properties around plant genetics—became the dominant structure of the industry (Graff, Rausser, and Small 2003). Patents can be held on plant varieties, transgenic traits, biotechnology processes, and research tools. Thus, the costs of negotiating access to the numerous component technologies are high (Srinivasan 2003). Further, a single company holding IPR over an important technology in the research and development chain can block the developments of other researchers or firms. Large firms quickly achieve market and proprietary powers that make it excessively difficult for smaller firms to compete or to enter the field. In the 1990s the largest firms absorbed over three-quarters of all enterprises engaged in biotechnology research (Moss 2013).

Out of the 1990s mergers and acquisitions surfaced an exceptionally consolidated, vertically and horizontally integrated industry. Six companies, colloquially referred to as the "Big Six," deserve most recognition, though 2017/2018 mergers and acquisitions have consolidated them even further to the "Big Four." While Monsanto may be most notorious, all have long histories in war and poison. Dow is the largest chemical company in the US. It manufactured nerve gases during World War II, knowingly allowed its pesticide product DBCP to cause permanent sterility in thousands of farmworkers, illegally dumped dioxins into Michigan's waterways for a century, and is heir to the Bhopal disaster (Doyle 2004). DuPont started as a gunpowder and explosives company, providing around half of those products used during the American Civil War and both world wars by Allied forces (Colby 1984). During peacetime DuPont diversified into chemicals; in their merging with Dow, the company became responsible for at least 238 Superfund sites (Reed 2018). Syngenta was formed through the merging of pharmaceutical giants Novartis and AstraZeneca's agrochemical lines. It is responsible for the infamous products paraquat, atrazine, and bee-killing neonicotinoids, as well as for the 1986 chemical disaster of the Rhine (Henry et al. 2012; Whitehorn et al. 2012). BASF and Bayer were both participants in IG Farben, dubbed the financial core of the Hitler regime and the primary supplier of chemicals used in Nazi extermination camps (Levy 1966). These are some of the same capitalist firms and family lineages that have profited the most from and have been integral to shaping the industrialization of food production through chemical, fossil-fuel-based inputs. While names have changed and companies have been spun off, dissolved, merged, and demerged, the crystallization of the Big Four seed-biotech companies—Bayer (Monsanto), Corteva (DowDuPont), ChemChina (Syngenta), and BASF—is a direct product of these legacies and actors.

Today the Big Four control nearly 60 percent of the commercial seed market globally, 75 percent of agrochemical sales, and 75 percent of private sector plant breeding research (Howard 2018). They control almost all genetically engineered seed sales. In contrast, in the 1960s around seventy basic pesticide manufacturers operated in the United States (Bryant et al. 2016), and in the 1980s the top nine seed companies commanded less than 13 percent of the seed market. Today, seed markets are most concentrated for commodity crops that are genetically engineered. Prior to 2017 mergers, Monsanto and DuPont alone controlled 66 percent of the US corn seed market and 62 percent of the soybean seed market (Matson, Tang, and Wynn 2012). Yet, even these high numbers belie the extent of market dominance. The same companies' patented seed varieties and transgenic traits are also licensed to the remaining retail market competitors, and the fewer than one hundred "independent" seed companies rely almost completely on the mega firms for licenses to genetic resources (Hubbard 2009). In the transgenic traits market, the biggest companies hold more than 95 percent of trait acres for corn, soybeans, and cotton globally (Moss 2013). Just one company—Bayer (Monsanto renamed)—has proprietary claim to transgenic traits in 95 percent of US soy and 82 percent of corn (Moss 2013). These two crops alone blanket over half of all US farmland. While seed markets are most consolidated for commodity crops, industry concentration extends to other crop types as well. Even prior to the 2017/2018 mergers, just three firms (Monsanto, Syngenta, and Vilmorin) controlled 60 percent of the global vegetable seed market (Hubbard 2019).

The agrochemical-seed-biotech industry employs a remarkable degree of political, technological, ideological, and market collaboration in order to maintain collective dominance. Companies work together both to create new frontiers of profit accumulation and to close those frontiers to potential competitors. For instance, joint ventures and cross-licensing agreements in research, development, and commercialization are used to maintain shared control. In 2012 Monsanto had cross-licensing agreements on proprietary germplasm and technologies with all other firms, Dow with four of the other five, and DuPont and Syngenta with at least three of the other companies (Shand 2012). With recent mergers, these cross-licensing agreements have only increased. Another industry standard for preserving collective market dominance is trait stacking. Stacking allows multiple firms to insert their patented traits into single crops and thus collaboratively enhance their oligopolistic reach through tie-in schemes, selling one product or service as a mandatory addition to the purchase of a different product or service. Longstanding rivalries between companies over patent infringement and antitrust claims have often been settled through cross-licensing agreements and technology sharing (Moss 2013). To the extent that there is competition within the agrochemical, seed, and agricultural biotechnology industries, it is almost entirely "oligopolistic rivalry, i.e., battles between monopoly-capitalist firms" (Foster, McChesney, and Jonna 2011).

The oligopolistic nature of the agrochemical-seed-biotech industry is not an isolated case. Capitalism inherently trends toward concentration of markets, wealth, and power. As a system driven by ceaseless competitive accumulation of profit, dominant firms seek to grow, reduce competition, and consolidate their market power. These are systemic compulsions of capitalism, not characteristics of any single industry. The past decades of neoliberal hyper-capitalist policy has rapidly intensified processes of consolidation and concentration. Today's most successful capitalist firms involve huge conglomerations of corporate power that are not only vertically and horizontally integrated within industries, but also include giant multinationals across many sectors (Foster, McChesney, and Jonna 2011). The agrochemical-seed-biotech industry is only one example of many, with conglomerates like Bayer operating 385 companies in eighty-three countries.

While capitalism's systemic drive is toward consolidation, governments do have the power to impose limits on monopolies. Consistent with the neoliberal agenda, over the past decades antitrust law has been almost entirely unenforced. Rather than curb market consolidation, the state has facilitated it by greenlighting monopolies. There has been virtually no limitation on agrochemical-seed-biotech industry mergers, even when anticompetitive behavior clearly exists. In court cases the US judiciary has consistently favored industry patent holders and the most stringent interpretation of their monopolistic property rights (Moss 2013). At the global level, the WTO and free trade agreements have rejected inclusion of mechanisms to limit corporate consolidation and anticompetitive behavior.

As a result of market concentration, seed prices have risen and market options have dwindled. In the US, between 1995 and 2008 corn seed prices increased 139 percent for GMO and 49 percent for non-GMO. Soybean seed prices increased nearly 200 percent for GMO and 96 percent for non-GMO (Matson, Tang, and Wynn 2012). As a whole, commodity seed prices doubled relative to prices received for crops (Moss 2013). In 2015, seeds were the third-largest cost for corn farmers following land rent and fertilizer (Bunge 2015). Technology fees for transgenic traits likewise have steadily risen, with Monsanto reportedly tripling its fee for soybean seeds from 2002 to 2008 (Hubbard 2009). Texas A&M researchers calculate that the Monsanto-Bayer merger will likely increase seed prices even further for US farmers (Bryant et al. 2016).

Lack of market choice outside of expensive Big Four technology has been manufactured through elimination of other options. In the US, it is increasingly hard to find non-GMO corn and soy seeds, or even GMO seeds with only one or two transgenic traits (Hubbard 2009). It has become difficult to purchase Bt seed (engineered with a gene from the bacterium *Bacillus thuringiensis* to produce insecticidal proteins) that is not also Roundup-resistant. Thus, what corn and soy seed is available on the market is largely technologically and legally bound to Bayer-Monsanto's herbicide (Hubbard 2009). While the industry asserts that market demand has driven GMO product development, evidence points more strongly to lack of alternative options in a highly monopolized market.

Though big, powerful, and oligopolistic, the Big Four are far from omnipresent. Around 80–90 percent of all seed used by farmers in the Global South remains outside of the so-called formal proprietary market (ETC Group 2013). The primary source of seed remains saved seed, including those exchanged through various local institutions and markets (ETC Group 2013). Further, while trends indicate steady growth of Big Four–patented GMO products, these continue to be concentrated in a handful of countries for the four major crops of soybean, corn, cotton, and canola. In 2019, around 190 million hectares of global cropland were planted in GMO commodity crops. Over 90 percent of this acreage is in just five countries—the US, Brazil, Argentina, Canada, and India (ISAAA 2019). The vast majority of genetically engineered crops are part of large-scale monocrop systems producing animal feed or automotive fuel, with a smaller amount refined into highly processed food fillers for direct human consumption.

Looking toward new frontiers of growth, the industry has more recently trended toward acquisitions and partnerships with companies based in the Global South, especially in India and Africa (ETC Group 2013). More emphasis is being put on fruits and vegetables, and especially crops for markets in the Global South. Compelled by the structural demands of capitalism, the Big Four will exert their accumulated wealth, power, and market dominance to amass more of it. In the process, they will further consolidate markets, as well as agricultural research and development. The consequences for the future of food and agriculture are disastrous.

Chapter 2

Science and Regulation in Service of Capital

As a small group of oligopolistic firms have come to dominate the agrochemical, seed, and biotechnology markets, their influence over agricultural research and development has expanded significantly. This has major consequences for the direction of the agrifood system. Driven solely by profit maximization, the oligopoly is entrenching a pesticide-intensive industrial-style agrifood system while undermining critically important agricultural research and innovation. The technologies being developed by the oligopoly are entirely premised upon the privatization of public science. The state itself has facilitated this transfer of public to private wealth. Moreover, the state has consistently worked to support already dominant corporations, extending their monopolies and private profits not only within the US but globally as well.

The Wedding of Public and Private Research

Private profit opportunities in agricultural biotechnology have rested entirely upon the exploitation of publicly funded science. Ideologically, the US champions the idea of laissez-faire, or "letting markets lead." In fact, though, the US has a long history of state developmental efforts in science and technology (Mazzucato 2013). With biotechnology, following the 1953 discovery of the structure of DNA, it was the state that advanced molecular biology by providing substantial public funding through the National Institutes of Health (NIH). Publicly funded basic sciences through the 1950s and 1960s led to breakthroughs in understanding the genetic code and figuring out how DNA replicates. Experiments in gene splicing were underway by 1967, with public funding paving the way for the emergence of genetic engineering in the 1970s. Gene splicing was achieved in 1973, launching technological possibilities for valuable new commodities. The NIH put a strong emphasis on advancing the technology of genetic engineering, both within their own labs and through grants that rose from two in 1975 to 1,061 in 1980 (Block 2008, 177). It was the US Treasury that financed the basic science leading to biotechnology and most of its breakthroughs, and has continued to pour funding into its applications. In general, biotechnology industries rely much more heavily on publicly funded science than other industries (McMillan, Narin, and Deeds 2000; Xia and Buccola 2005).

The persistent rhetoric of private sector leadership in technological innovation masks the importance of public research and lends support to its privatization. This is the case not only for biotechnology but also for technology ranging from aviation and nuclear energy to computers and the internet (Mazzucato 2013). Privatization of technology and information has been facilitated especially by policies implemented during the presidencies of Ronald Reagan and George H. W. Bush. In the 1980s, a range of initiatives were instituted that encouraged public-private partnerships, technology transfer to private industry from public institutions, and public subsidies to technology companies (Block 2008, 180). Such policies, which continue to this day, are tied to wider neoliberal shifts that increasingly subject public universities, research, and science to the logic of the market. Public funding for universities has been cut as they have become dependent on industry for funding, public research has been forced to prioritize commercializable products, intellectual property rights have intensified, and the logic of business has been applied to all realms of universities (Lave, Mirowski, and Randalls 2010; Newfield 2008; Lotter 2008).

The Bayh-Dole Act of 1980 was especially significant to agricultural biotechnology and the rise of the agrochemical-seed-biotech oligopoly. The act allowed public universities to privatize and patent the results of research funded by public monies. Prior to Bayh-Dole, federally funded NIH research was in the public domain and could be adopted by any researcher, allowing for proliferation of innovation (Garcia-Sancho 2012). Bayh-Dole instead permitted publicly funded research to become the sole property of single individuals, institutions, or firms. Subsequently, the Federal Technology Transfer Act of 1986, lobbied for by Monsanto, mandated federal agencies to make technology transfer to private industry part of their mission (Hauter 2012; Block 2008). Universities adapted by privatizing and selling their technologies, encouraging direct industry investment, and establishing special technology transfer offices staffed with professional marketers (Feller 1990). The rate of growth in annual patent filings rose tenfold in the two decades following Bayh-Dole (Jasanoff 2005). Perhaps more than any other sector, biotechnology showed significant potential for profitable commercialization (Argyres and Liebeskind 1998, 428).

Though many scientists were initially "scandalized" by their NIH-funded colleagues' early commercial biotech projects, a turn to the private sector rapidly became the norm (Block 2008, 177).

The role of public land-grant universities in this trend toward corporatization and privatization is most concerning. Since their establishment in 1862, land-grant universities have played a critical role in agriculture, including in plant breeding, responding to agricultural challenges, and pioneering research on environmental stewardship. Much research has historically been done in partnership with farmers and distributed broadly through rural extension offices. In addition, high-quality seed has long been made available to producers on a non-proprietary basis (Matson, Tang, and Wynn 2012). With 1980s policy changes, public research dollars have shifted away from these public goods and toward advancing what are perceived as cutting-edge market growth sectors. This has resulted, for example, in a substantial increase in funds, faculty, and students dedicated to agricultural biotechnology, alongside a decline in public plant breeders (Hess 1991). With neoliberal shifts in logic and policy, public land-grant universities are increasingly acting as businesses themselves. They have joined Monsanto in programs that sue farmers for seed patent infringements, including using private investigators to catch farmers in the act of seed saving (FWW 2012). They have also intervened on the side of Monsanto in high-profile court disputes, arguing that strict intellectual property rights around seeds are important for their own contributions to innovation. The University of California alone spends $28 million annually employing lawyers to manage its IPR portfolio (Bond-Graham 2013).

Such changes in priorities correspond with the shrinking of public funding for agricultural research and a dramatic rise in the role of private capital. With the advent of biotechnology, in the early 1990s industry funding of agricultural research at land-grant universities surpassed USDA funding. In 2009, industry investments were $822 million as compared to the US Department of Agriculture's $645 million (FWW 2012). The direction, nature, and outcome of scientific research is significantly distorted by this private sector involvement (Mirowski 2011; Newfield 2008). Independent evaluation of agricultural technologies has been severely limited by corporate influence, intellectual property rights, and lack of funding for research that might prove unfavorable to industry (Krimsky et al. 1996; Hilbeck et al. 2015; FWW 2012). Nearly half of land-grant agricultural scientists surveyed in 2005 stated that they had received funding from a private company. One-third reported consulting for private industry (Goldberger et al. 2005). The outcomes of research are significantly influenced by funders, including frequent imposition of strict prepublication review (Mirowski 2011; Bowring 2003; Campbell, Louis, and Blumenthal 1998; Parfitt 2013).

In addition to financial dependence, public university agricultural scientists are often reliant on technical cooperation from seed companies. As a Cornell professor of entomology told the *New York Times*: "People are afraid of being blacklisted . . . If your sole job is to work on corn insects and you need the latest corn varieties and the companies decide not to give it to you, you can't do your job" (Pollack 2009). Research involving agrochemical-seed-biotech industry products is generally performed only with company approval. As another scientist told the *New York Times*, "If a company can control the research that appears in the public domain, they can reduce the potential negatives that can come out of any research" (Pollack 2009). The agrochemical-seed-biotech industry is notorious for harassing and directly attacking the credibility of researchers who raise concerns with their products ("The Biologist Who Challenged Agribusiness" 2016).

Most generally, private sector spending on agricultural research has risen dramatically while public sector spending has decreased. Private agricultural research expenditure in the US increased by 1,300 percent between 1960 and 1996. During this same time, public spending stagnated and more recently has decreased (Matson, Tang, and Wynn 2012). In 2006, private research and development funding stood at $7.4 billion in comparison to $5.7 billion in total public funding (FWW 2012). By 2010, the private sector was spending $11 billion on agricultural research, with the most rapid growth in commodity crop seeds and biotechnology traits (Knutson 2013). Globally, public funding for agricultural research and development has followed the same trend, stalling by the 1990s while private sector research increased (Paul and Steinbrecher 2003). Of most consequence, this private sector research is exceptionally concentrated and dictated by the profit interests of the agrochemical oligopoly. The four dominant firms account for more than three-quarters of all private sector research and development in both seeds and agrochemicals globally (Shand 2012). Their research dominance has a remarkable influence on the evolution of global agriculture.

Profit-Driven Technology

On average, the Big Four devote upward of 70 percent of seed and crop research to genetic engineering (Shand 2012). Despite promises of drought-resistant, nutritionally enhanced, and other wonder crops that are always right around the corner, more than forty years into research and twenty-five years into commercialization the industry's only substantial biotech products remain crops engineered to resist herbicides or to produce their own insecticide (*Bacillus thuringiensis*). Four crops (soy, corn, cotton, and canola) engineered with one or both of these transgenic traits account for 99 percent of global acres planted in GMO crops (ISAAA 2019). The other less than 1 percent of worldwide GMO acres are in alfalfa (1.3 million hectares), sugar beets (473,000 hectares), sugarcane (20,000 hectares), papaya (12,000 hectares), safflower (3,500 hectares), potatoes (2,265 hectares), eggplant (1,931 hectares), and a very small amount of squash, apples, and pineapple (ISAAA 2019). Most telling, 88 percent of all acreage devoted to GMO crops contain a trait for resistance to herbicide also manufactured by the agrochemical-seed-biotech industry (ISAAA 2019). The primary trait in cultivation remains resistance to Bayer-Monsanto's Roundup (glyphosate). These crops led to a 527 million–pound increase in herbicide use between 1996 and 2011 just in the United States (Benbrook 2012).

The immediate horizon of new biotech crops includes more of the same pesticide+seed combinations in increasingly toxic versions. Put simply, this is because herbicide-resistant crops speed up the pesticide treadmill in monocrop agriculture. Herbicide-resistant crops allow for spraying both before and during the growing season without harming crops. As larger swaths of land have been brought into production with single crops and single herbicides year after year, weeds have evolved resistance (International Survey of Herbicide Resistant Weeds 2013; Gurian-Sherman and Mellon 2014). Glyphosate-resistant "superweeds" are found today on nearly 100 million acres of farmland across the US (Landrigan and Benbrook 2015). The industry's response has been to engineer crops to withstand more toxic herbicides and combinations of multiple herbicides.

After years of controversy, in 2017 Dow announced the launch of crops engineered to be resistant to both 2,4-D and glyphosate. Branded Enlist Duo, the new GMOs were explicitly pitched by Dow as the answer to glyphosate-resistant superweeds (Dow AgroSciences 2011). Scientists predict that the new crops "will facilitate a significant increase in herbicide use" as weeds develop resistance to multiple herbicides and farmers respond with the most readily available chemical technologies (Mortensen et al. 2012, 75). Dow and Monsanto themselves predicted a surge in herbicide use, while the United States Department of Agriculture (USDA) estimated a three- to sevenfold increase in agricultural use of 2,4-D within three years, from 26 million to as much as 176 million pounds per year (USDA 2013). The herbicide 2,4-D is a component of Agent Orange, which was used as defoliant in the Vietnam War and linked to Parkinson's disease, non-Hodgkin's lymphoma, and reproductive problems. Glyphosate was classified as a "probable carcinogen" by the World Health Organization in 2015. They have not been tested in combination for health or environmental impacts. The Environmental Protection Agency's own assessment found that Enlist Duo is highly toxic to many plant and animal species. After it was initially approved, the Environmental Protection Agency (EPA) under Barack Obama's administration revoked registration of Enlist Duo due to litigation by a coalition of environmental groups. The decision was quickly reversed with Scott Pruitt at the helm of Donald Trump's EPA. Pruitt reaffirmed the EPA's initial approval and then dramatically expanded it, greenlighting Enlist Duo to be used in more than thirty-four states and on cotton in addition to corn and soybeans.

The EPA and USDA also approved commercialization of Monsanto's dicamba- and glyphosate-resistant soybeans and cotton for planting in 2017. Dicamba has been linked to elevated risk of numerous cancers (Lerro et al. 2020). In the past, dicamba was seldom used on farms during the warm summer months because it is highly prone to volatizing, or turning to a gas and migrating to nearby fields. Monsanto claimed it had solved this problem with a formulation called VaporGrip that would make dicamba less volatile. By the end of their first growing season, a reported 3.6 million acres of soybeans in more than twenty states were damaged by dicamba that had drifted from neighboring fields (Philpott 2018b). Millions more acres of shrubs, trees, vegetables, and lawns were likely impacted. Kevin Bradley, a University of Missouri weed scientist, told *Mother Jones* journalist Tom Philpott: "In my opinion we have never seen anything like this before; this is not like the introduction of Roundup Ready or any other new trait or technology in our agricultural history" (Philpott 2018b). Despite millions of acres in crop damage in 2017, 2018, and 2019, in 2020 dicamba-resistant crops were planted on an estimated 60 million acres. Farmers report having to buy the dicamba-resistant seeds just to protect their crops from drift (Philpott 2018a). Monsanto estimated that dicamba use would increase 88-fold for soybeans and 14-fold for cotton with the release of their new herbicide-resistant seeds. In 2016 the company announced a nearly $1 billion project to expand its dicamba plant in Louisiana, telling investors that dicamba-resistant soybeans would eventually be planted on 80 million acres across the US and generate $400 million annually (Philpott 2018a). In 2020 Bayer-Monsanto petitioned the USDA to approve a new corn engineered to tolerate dicamba along with four other pesticides—glufosinate, quizalofop, 2,4-D, and glyphosate—at the same time.

A majority of the genetically engineered crops being considered for deregulation by the USDA are developed for herbicide resistance. Dow has a patent claim on a mechanism that would allow nine types of herbicide resistance to be engineered into a single plant (Keim 2012). In other words, a single crop could be blanketed with nine different herbicides without being impacted. Monsanto's failed 2015 bid for Syngenta (which held the largest share of agrochemicals) and its subsequent successful turn to chemical company Bayer made clear that pesticide+seed combos are the industry's priority technological innovation. Not only do these product innovations deepen and speed up a toxic and anti-ecological pesticide treadmill, but they are treading toward the last line of defense in chemical agriculture. This could lead to massive disruption in food production systems as chemicals lose their effectiveness and "super pests" become increasingly prevalent (Benbrook 2001, 206). The industry's strategy of simply ramping up chemical resistance in the face of nature's evolution is indicative of the logic of capital. In ecological terms it is a a dead-end, potentially catastrophic strategy. But it is absolutely logical to the goal of short-term profit gain divorced from public costs and wider impacts.

In contrast to the primary GMO crops, which are herbicide-resistant, secondary GMO Bt crops were generally believed to have reduced insecticide application in the United States. Bt crops produce their own *Bacillus thuringiensis* insecticide and therefore require less applied insecticide (Benbrook 2012). More recent analysis, however, has shown that while applied insecticide has decreased with the introduction of Bt crops, neonicotinoid (also an insecticide) seed treatment has risen dramatically, to an extent that insecticide use almost tripled in overall corn hectares by 2011 (Douglas and Tooker 2015). Insecticidal seed treatments are typically one component of larger packages of technologies sold to farmers that, for instance, "can include germplasm (i.e., crop variety), up to eight transgenes, and up to six or more different seed treatments (fungicides, nematicides, and insecticides)" (Douglas and Tooker 2015, 5092). Like the evolution of herbicide-resistant weeds, Bt crops have also created new challenges with the evolution of Bt-resistant pests (Gassmann et al. 2011; Dhurua and Gujar 2011).

Most fundamentally, the problem is with the monocrop model of agriculture encouraged and extended by both herbicide-resistant and Bt crops. The industry's biotechnology products rely on the industrial, chemically dependent monoculture model of agriculture and cement this model by simplifying industrialized production further (Benbrook 2001; Hilbeck 2008). Besides herbicide resistance and Bt, other traits that have dominated agricultural biotechnology research—including transportability, shelf life, visual appeal, suitability for food processing, and uniformity—are all in the vein of forcing crops, environments, and agricultural communities into chains of highly mechanized capitalist production (Bowring 2003). This narrow technological pathway is not necessarily innate to genetic engineering but to its deployment by dominant capitalist firms within structures of competitive profit-making. How or if genetically engineered crops could be used instead in more biodiverse and low-input systems remains largely theoretical.

Profit opportunities in seed and crop research are almost entirely reduced to monocropped commodities that form the foundation of the industrial food system. The dominant capitalist seed firms invest nearly half of their research dollars into a single crop: corn. Almost all of their research funding is on less than a dozen crops (Mooney 2015). While human practices of commoning have led to the development of around 7,000 domesticated species—essential to biodiversity and resilience of the agrifood system—seed companies are interested in only 150. Orphaned, open-pollinated varieties have been neglected because there is no profit to be made; meanwhile, billions are invested into a handful of commodity crops (Zerbe 2015). The profit-driven research priorities of the private sector also significantly distort public sector spending. Very little research into potential crop improvements has been done on the small-grain cereals, tubers, and legumes that are cultivated by hundreds of millions of smallholding farmers (Kiers et al. 2008).

Agricultural science and technology dominated by monopolistic firms and profit-making opportunities is thwarting wider innovation and marginalizing research that is essential to environmental sustainability. It is narrowing diversity of research, diversity in fields, and global genetic diversity. Incentives for capitalist firms to innovate will always be narrowly defined by returns. Here arises a most critical contradiction in the productivity and innovativeness of capitalism. While capitalism is largely celebrated for being a system that encourages innovation and technological progress, privatization of knowledge and information is increasingly a fetter on their advancement. As the agrochemical-seed-biotech industry has consolidated its grip on intellectual property, innovation in agricultural biotechnology has declined (Moss 2009; Fernandez-Cornejo and Schimmelpfennig 2004; Schimmelpfennig, Pray, and Brennan 2004). More than this, it has limited innovation in other, critically important research and development pathways. With the agrochemical industry playing a defining role in agricultural research and development and public science increasingly embedded in capitalist market demands, there is an abandonment of science and technology designed for purposes of broader social benefit (Mirowski 2011).

The technological trajectories that are being entrenched, as others are marginalized, are proving increasingly perilous. There is maturing scientific consensus regarding the need to radically transform agrifood systems for long-term sustainability (IAASTD 2009; UNCTAD 2013; Molden 2007; Kiers et al. 2008). The most comprehensive global assessment of agriculture to date, the

International Assessment of Agricultural Knowledge, Science and Technology for Development (IAASTD), strongly concluded that the world must fundamentally "reorient our food and agricultural systems toward sustainability, health, bio-cultural diversity, ecological resilience and equity" (Ishii-Eiteman 2009, 1). The IAASTD emphasized agroecological methods, increasing farm diversification, reducing agrochemical inputs, and enhancing biodiversity conservation. These are precisely the pathways being marginalized and undermined by the agrochemical-seed-biotech industry's market and research dominance. Production systems—especially of life's most basic necessity—cannot be sustainable, democratic, or equitable when subsumed by the logics of privatization and competitive profit accumulation. Vast monocrops engineered by way of privatized science to sustain heavier doses of pesticides are merely one illustration of this. The technologies that are being put to the service of private profit accumulation by the largest corporations are not the problem in and of themselves. It is the very drive at the core of the system that must be called into question.

Regulation by and for Capital

The drives of capitalism do not sustain themselves on their own. The idea that capitalist markets can be free of government intervention is an ideological ploy, having no basis in how the world actually operates. The state has always been, and remains, vital to maintaining a capitalist system. The state is the sole institution with the authority to define and enforce what counts as legal. With this authority, the capitalist state manages a system of private property, markets, social order, and ideological reproduction, including backing processes of expropriation of commons (Harvey 2005; Wood 2003). There is no part of the capitalist accumulation process that is not dependent upon the legal, political, and administrative order supplied by the state (Gindin et al. 2011; Wood 2003).

The state also has the authority to regulate capital and redistribute wealth that trends toward the top in a capitalist system. Thus, it is a critical site of struggle. Even under capitalism, the state can expand or restrain democracy, equality, and justice. It can be the enforcer of the "rights" of the capitalist elite to exploit, or it can actively work to lessen the hierarchies and inequalities inherent to the system. While capitalism is intrinsically antidemocratic, exploitative, and class based, the state can at least alleviate its merciless extremes. How a state functions in this regard is neither monolithic nor predetermined. States are a "malleable product of human interactions," a space where divergent interests compete and interact (Gilbert 2013, 10).

Understanding the rise of the agrochemical-seed-biotech oligopoly in relation to states requires a brief detour to wider historical context. Since the Cold War especially, capitalist competition is increasingly structured around global markets. Transnational corporations shop around for the best tax benefits, the highest subsidies, the lowest environmental and labor standards, and the strongest property rights (McCarthy 2004). Circuits of production, accumulation, and finance are largely transnational (Robinson 2007). For all of these globalizing tendencies, what is colloquially referred to as globalization has also rendered states and their capacities more, not less, relevant. Ellen Wood contends that the state has been "an indispensable instrument in the process of spreading capitalist imperatives, not only in the sense that the military power of European nation states has carried the dominating force of capital to every corner of the world, but also in the sense that nation states have been the conduits of capitalism at the receiving end too" (2003, 22). As the most reliable ensurer of the conditions necessary for capitalist accumulation, the state is more essential than ever to global capitalism. While the past decades' global neoliberal project has been ideologically opposed to government, in practice it has been characterized by consistent state intervention to subject ever more facets of life to market relations (Panitch and Gindin 2013; Brown and Shenk 2015).

The globalizing tendencies of capitalism have especially been extended by way of imperial domination. American empire and its allies have worked to establish globalized capitalism, act on behalf of global capital, and dominate less powerful states through capitalist institutions (Gindin et al. 2011, 109). While global hierarchy is ultimately enforced by the US military, the main feature of today's capitalist imperialism is that it operates as much as possible via economic imperatives. Debt, the rules of trade, foreign aid, and the entirety of the financial system are tools of manipulation for ensuring that economies serve the interests of the dominant global economic powers (Wood 2003, 134). It is not only the US or its elite that are the benefactors of the new imperialism. Elite classes of many nations are critical contributors and beneficiaries (Panitch, Henwood, and Lilley 2011). Imperial domination takes place through capitalist institutions and imperatives, and global capitalism itself has been intensified by way of imperial state power.

After World War II, the economic hegemony of the US and its (shifting) allies was asserted through the Bretton Woods system, the International Monetary Fund (IMF), and the World Bank, and later through the General Agreement on Tariffs and Trade (GATT). Bretton Woods was replaced in the early 1970s with the Washington Consensus and neoliberal structural adjustment as mediated through the likes of the IMF and the World Bank. Structural adjustment made subordinate states more vulnerable to the pressures of global capital by liberalizing economies, privatizing public services, changing regulations on finance, and gutting social protections and supports.

Also fundamental was that in 1994 the World Trade Organization (WTO) was created out of the GATT with legislative and judicial powers that enable states to challenge other states' laws in the interests of free trade. Following the creation of the WTO, free trade agreements have similarly worked to free capital of regulatory regimes that restrict profit maximization (labor and environmental protections, for instance), while extending the regulatory forms that support private property and dominant corporations. While often referred to as promoting deregulation, they are in fact regulation in a particular form. They are highly bureaucratic treaties that, far from being about encouraging more open trade, in practice typically function to control trading conditions to benefit dominant corporations and nation states. Rife with contradiction, free trade policy has been central to increasing the dominance of transnational corporations and international oligopoly groups (Foster, McChesney, and Jonna 2011). One oligopoly group that has benefited mightily from free trade and wider neoliberal, hypercapitalist policy is the agrochemical-seed-biotech industry. The industry has been a notable priority in the United States' imperialist commitment to the expansion and intensification of capitalism.

The United States' preferential treatment of already dominant agrochemical-seed-biotech companies is entwined with corporate influence over the state, export interests, and dominant ideas about high technology and economic competitiveness. Much state activity related to biotechnology in the US—from nonregulation of genetic engineering, to patent rights, to government funding of technology—has unfolded within a dominant political discourse around national competitiveness in the global market (Azmanova 2015). From the 1970s and 1980s, high-technology sectors have come to occupy a privileged space in conceptualizations of national economic competitiveness. The interests of particular high-tech corporations and sectors are projected as benefiting economies and society as a whole. From its inception, agricultural biotechnology was represented in the US as a key component of the "knowledge economy," and thus a major driver of growth, progress, and prosperity (Newell and Glover 2003). Significant debate, including around the safety of food products, was and continues to be framed around maintaining a competitive lead in biotechnology (Jasanoff 2005). In this, the state's role is constructed as mere promotion of biotechnology to benefit "the economy," without asking critical questions about who is benefiting and how, and at what and whose expense.

Every federal administration since Ronald Reagan's has been willing to act definitively on behalf of the biotechnology industry (Newell 2009, 40). The administration was pioneering in its demolition of regulations and strong hand in pushing the industry forward. In 1986 it issued the Coordinated Framework for the Regulation of Biotechnology, solidifying a government consensus that no new laws would be passed and that biotechnology products would not be treated any differently from products that did not involve genetic engineering (Krimsky 2005). The process of genetic engineering was rendered safe in law, with only the products in need of narrow risk assessment under already existing regulations (Jasanoff 2005, 2014). Genetically engineered products were treated as novel enough to secure strong intellectual property rights, but simultaneously "substantially equivalent" to other products in regard to environmental, health, safety, and other regulatory protections. Regulation related to biotechnology was designed around extreme pro-corporate, anti-regulatory ideology as opposed to any kind of coherent scientific or legal rationale.

The dominant chemical corporations that seized agricultural biotechnology largely shaped its regulatory regimes. Often this was in direct competition with the interests of less powerful biotechnology firms. Long before biotechnology, major conglomerates like Monsanto and DuPont had well established histories asserting their interests upon the state (Colby 1984; Newell and Glover 2003, 4). Most notable because of their pioneering role, influence, and sustaining dominance in agrochemical-seed-biotech endeavors is Monsanto. Monsanto lobbied avidly for the "coordinated framework" issued by the Reagan administration. Monsanto's position was actually in conflict with ideologically pure conservative government officials and some smaller firms that opposed any regulation at all. Monsanto executives were weary of battles with environmentalists through the 1970s and carefully monitoring growing public controversy around biotechnology. They wanted to shape regulation that would reassure the public of government oversight but without actually impinging on their endeavors (Charles 2001). They desired that no new laws or debate be opened up by Congress, but instead a veneer of regulation that guaranteed their products would be commercialized as long as they fulfilled a clear list of requirements (Charles 2001, 28). Against less powerful biotech firms and official Reagan doctrine, Monsanto got what it wanted in the coordinated framework. It was adopted "almost verbatim" from an industry proposal written by Michael Taylor, who from the Reagan administration through to the Obama administration has moved back and forth between working for and "regulating" Monsanto (Hauter 2012, 247).

The Reagan administration's coordinated framework left it to federal agencies to apply existing statutes to oversight of biotechnology products. As Monsanto's products moved closer to commercialization, the company worked closely with the George H. W. Bush White House to ensure reforms that would ease the process of bringing new products to market (Eichenwald, Kolata, and Petersen 2001). This time Monsanto's wishes were in contest with smaller biotechnology companies like GMO tomato developer Calgene. Calgene's chair insisted that the American people wanted credible federal oversight and that transparency and rigorous pre-market testing would be better for the development of the industry (Charles 2001, 134). Monsanto again had its way with Michael Taylor back working within the Bush administration and top executives paying multiple visits to President Bush himself. Against the wishes of many of its own scientists, the Food and Drug Administration (FDA) enshrined a policy of leaving it up to companies to conduct pre-market testing of food products and voluntarily provide information confirming their safety (Krimsky 2005). A *New York Times* piece from 2001 summarizes what remains the case today: "[T]he White House complied, working behind the scenes to help Monsanto—long a political power with deep connection in Washington—get the regulation that it wanted. It was an outcome that would be repeated again and again . . . If the company's strategy demanded regulations, rules favored by the industry were adopted. And when the company abruptly decided that it needed to throw off the regulations and speed its foods to market, the White House quickly ushered through an unusually generous policy of self-policing" (Eichenwald, Kolata, and Petersen 2001).

These early stories in the regulation of agricultural and food biotechnology are only a glimpse of the extensive relationships between agrochemical-seed-biotech firms and the US state. The list of revolving-door relationships, for instance, is long. Between acting as US secretary of defense under Presidents Gerald Ford and George W. Bush, Donald Rumsfeld ran the company that produces NutraSweet, which was purchased by Monsanto (Hauter 2012). In President Bill Clinton's administration, Mickey Kantor bounced between trade negotiator and Monsanto's board (Charles 2001, 166; Eichenwald, Kolata, and Petersen 2001). As NPR journalist Dan Charles writes in his book *Lords of the Harvest*, Monsanto executives have typically had "good friends" in the White House (Charles 2001, 166). Monsanto is not distinct among the agrochemical-seed-biotech oligopoly. In 2015, of DuPont's nineteen in-house lobbyists, eleven came straight out of government. Around the same time, Dow boasted nineteen of thirty-three revolving-door lobbyists (" 'Big 6' Pesticide and GMO Corporations" 2015). These are only the most visible relationships. The elite social networks that bring big business and regulators together run significantly deeper and wider (Newell 2009).

The positions and interests of the US government and corporations have influence far beyond American borders. The US government consistently champions the industry's interests globally, working to actively promote its technologies and weaken regulations that constrain its profits and monopolization. As described, the relationship between the US state and the agrochemical-seed-biotech oligopoly requires understanding the broader operation of the US as a "capitalist empire" devoted to deepening and expanding capitalist logic globally (Panitch and Gindin 2013). At the same time, support of the industry goes beyond mere promotion of capitalism. As a result of decisions that positioned the US early as the epicenter of agricultural biotechnology, the US now has a major interest in ensuring that the rest of the world embraces it. The US is the world's largest agricultural exporter, with around half of its value derived from soybeans, cotton, corn, and wheat and their processed products. Besides wheat, these crops are almost entirely genetically engineered. It is not only commodity growers and the biotechnology-seed industry that are invested in maintaining the stability and profitability of status quo production; it is also grain traders, the meat industry, food processors, retailers, fertilizer and machinery companies, and banks, all of which are also highly oligopolistic and exert tremendous influence on the state to maintain and extend the industrialized food system (Newell 2009).

When the WikiLeaks release of US State Department cables exploded in 2010, nearly one thousand cables revealed the extent to which the US collaborates with industry to influence other governments' policies regarding agricultural biotechnology. According to analysis by Food and Water Watch, every US diplomatic post worldwide is instructed to "pursue an active biotech agenda" (FWW 2013, 3). Tactics range from drafting GMO crop approval legislation for ostensibly independent countries, to sponsoring "seeing is believing" GMO excursions targeted at journalists and diplomats, to distributing curriculum in high school classrooms of foreign countries (FWW 2013; Schnurr 2013). Strategies and goals have been designed and implemented in partnership with Monsanto, DuPont, Syngenta, and other agribusiness corporations. The US government reproduces the industry's highly contested and sometimes outright false claims, including that GMOs reduce pesticide use, increase yields, benefit the world's poorest farmers, are important to mitigating climate change, and are necessary to feed the world. The WikiLeaks cables also divulged how the US has directly intervened in other countries' affairs on Monsanto's behalf to negotiate seed royalty settlements, accelerate approval of their crops, and extend patent lengths (FWW 2013).

None of these revelations were particularly surprising or entirely unknown. The US has long worked aggressively to weaken other governments' regulatory oversight of genetically engineered crops and restrictions on imports (Newell 2008, 2009). When, for example, Egypt demanded in 1998 that GMO food imported from the US be labeled, the US responded by threatening to ban all trade in soy and maize between the two countries (Newell and Glover 2003). The US forcefully uses international trade rules to attack laws, including bringing a case before the WTO against the European Union's more precautionary stance on genetically engineered crops. At other times the US has used the mere threat of bringing a WTO case in order to force less powerful countries to conform to desired biotech policy. WTO rules and free trade agreements are increasingly being used in efforts to "harmonize" GMO-related laws to align with US anti-regulatory legal regimes (Newell and Glover 2003; Hansen-Kuhn 2016).

The case is similar and overlapping with agrochemicals. Again, the US uses binding trade rules and other economic and political pressure to advance industry and export interests. When Thailand proposed a ban of glyphosate in 2019, the US government worked directly with Bayer-Monsanto to kill the proposed policy by threatening trade disruption. The following year, Mexican president Andrés Manuel López Obrador called for the phasing out of both glyphosate and Bayer-Monsanto's genetically engineered glyphosate-resistant corn. Once again, industry is working hand in hand with the office of the US Trade Representative and other US agencies to interfere through trade agreements and other means (Gillam 2021). These are just two recent examples of many. In Europe, negotiations over the proposed Transatlantic Trade and Investment Partnership (TTIP) indicate a push for "regulatory convergence" that would weaken EU pesticide regulation, allow higher levels of pesticides on foods imported from the US, and block public access to "confidential business information" about pesticide ingredients and potential dangers (Smith, Azoulay, and Tuncak 2015). The US currently allows use of 82 pesticides that are banned in the European Union, and the industry is openly concerned about the precedent of the precautionary principle (Smith, Azoulay, and Tuncak 2015). Through every administration, the pattern of interference in other countries' laws to promote agrochemicals alongside the industry is consistent.

The agrochemical-seed-biotech oligopoly relies entirely upon states to secure monopoly intellectual property rights, fund public science that it privately profits from, and advance its interests through a variety of policies. Most centrally, the US federal government has played an essential role creating the conditions for the emergence and consolidation of the oligopoly. Through imperial mechanisms it forces the industry's products on sovereign nations while at the same time intensifying global monopoly capitalism generally. In circular fashion, as the industry has consolidated wealth and power, it has used that power to decidedly influence American policy. The agrochemical industry's occupation of Hawai'i is not unrelated to these global processes. The same forces—imperialism, capitalist empire, and expropriation of commons—underlie the occupation of the most geographically isolated islands on the planet by the largest chemical corporations in the world.

Chapter 3

"GMO Ground Zero"

Hawai'i has long been a place of exploit for extraction of wealth by capitalist interests and, relatedly, a hub of US imperial maneuverings. Today, capital and empire take shape around a dominant transnational corporate tourism economy and extensive US military control of land and sea. At the edges of this tourism- and military-based economy and in the tracts of abandoned sugar and pineapple plantations, the agrochemical-seed-biotech industry has grown to be the largest agribusiness in the islands. This chapter outlines the nature of the industry's operations in Hawai'i, including which companies are involved, what they do, where they are located, and the impacts of their intensive pesticide use. Following chapters more deeply examine the question of why Hawai'i has come to play such a central role in the industry's global chains of production and to whom the benefits and burdens of their occupation accrue.

Chemical+Seed Operations in the Islands

Precolonial Hawaiian agriculture was integrated with human settlement on an estimated 380,000 acres. Sophisticated agroecological systems sustained high-density populations that may have exceeded well over half a million people (Lincoln and Vitousek 2017). In contrast, contemporary colonial-capitalist Hawai'i imports nearly 90 percent of its food and exports an estimated 79 percent of all agricultural production (Melrose, Perroy, and Cares 2016). The State of Hawaii's most current 2015 agricultural land use survey reports a total of 151,830 acres in active crops and 761,430 acres in pasture. Primary crops include genetically engineered seeds, macadamia nuts, coffee, forestry, flowers, papaya, cattle, pineapple, and basil, sustained by the labor of primarily first- and second-generation migrants who make up the majority of farmworkers (Melrose, Perroy, and Cares 2016).

Seed-biotechnology research and development is Hawai'i's largest agricultural industry in terms of acreage occupied. The seed industry first arrived to

Hawai'i in the 1960s, operating on the fringes of plantation lands and primarily involved in hybrid corn endeavors (Brewbaker 2003). James Brewbaker, a plant breeder at the University of Hawai'i's College of Tropical Agriculture and Human Resources (CTAHR), demonstrated the ability to grow corn in the islands year-round with fewer weather challenges than other primary locations like the Midwest, Puerto Rico, and South Florida. Brewbaker encouraged "seedmen" to come to the islands, where they gained small leases but were shut out from prime agricultural lands dominated by sugar (Voosen 2011). In 1969, seed companies were located on around 41 acres on Kaua'i, 120 on Maui, and 430 on Moloka'i (Freese, Lukens, and Anjomshoaa 2015).

During the megamergers of the 1990s, seed companies operating in Hawai'i were either acquired by chemical-pharmaceutical corporations or went out of business (see chapter 1). Corn States was bought by Monsanto, Illinois Foundation Seeds by Dow, Trojan Seed Co. by Pfizer and then Monsanto, Funk's G by Ciba and then Cargill and eventually Monsanto, Northrup-King by Syngenta, and Pioneer Hi-Bred by DuPont. Chemical corporations' takeovers of seed companies were entirely linked to emerging profit opportunities in agricultural biotechnology. As they bought up seed companies operating in Hawai'i, some of the earliest GMO field trials were conducted in the islands. All of this took place at the same moment that sugar plantations were shutting down and the State of Hawaii and landowners were searching for the next agricultural industry to take its place. As agricultural biotechnology, chemicals, and seeds became a single consolidated industry with tentacles already in Hawai'i, space was being made on larger tracts of land equipped with water infrastructure and near communities left jobless by closing plantations.

Today, Hawai'i is at the epicenter of the agrochemical-seed-biotech oligopoly's global chains of production. Until very recently, it was estimated that nearly every genetically engineered corn seed grown globally has touched Hawai'i in its development (Pollack 2013). The process of inserting traits into plants begins in laboratories in the continental United States. Plants are then transferred to Hawai'i, where traits are bred into seed varieties over several generations, going through many planting cycles over more than three years (Schrager 2014). Three or four crops can be grown per year in Hawai'i, dramatically accelerating the pace of production and profit generation. Some amount of parent seed increase and grow-out also occurs in Hawai'i. Once developed, parent seed is shipped to places like Puerto Rico and Argentina for mass multiplication before being distributed to farmers in North and South America for planting. As stated by a Syngenta spokesperson, "these seeds travel all over the Mainland, and sometimes South America and Europe before the farmer ever gets it" (Lyte 2015). Seed is the State of Hawaii's largest agricultural commodity. Seed exports reached a high of 12 million pounds in 2009–2010, but dropped significantly to 3.8 million pounds in 2019–2020 (USDA National Agricultural Statistics Service 2020). With global restructuring of the industry and local resistance, it appears that parent seed production has been largely relocated just in the past couple of years.

Hawai'i has also consistently hosted more experimental genetically engineered crop field tests than anywhere else in the United States until recent years. Nearly 3,600 permits have been issued for field tests since 1988 (USDA 2020). In 2015, 154 different field tests were conducted at 1,228 sites. These numbers dropped every year from 2015 to 2018. In 2017 and 2018, Hawai'i for the first time had fewer test sites than Puerto Rico, which has historically hosted the second highest number of GMO field tests in the United States (USDA 2020). In Hawai'i, more than 90 percent of GMO field tests are corn and soy, with herbicide resistance by far the most frequently tested trait (Freese, Lukens, and Anjomshoaa 2015; ISB 2015). The large agrochemical-seed-biotech corporations conduct nearly all GMO field tests, while public sector institutions have been responsible for only around 1 percent in recent years (Freese, Lukens, and Anjomshoaa 2015).

The companies involved in seed-biotech research and production in Hawai'i are primarily the companies profiled in previous chapters that form a global oligopoly. Up until 2016, these were five of the Big Six—Monsanto, Dow, DuPont, Syngenta, and BASF. Mergers and acquisitions of 2017/2018 shifted ownership and name branding within the oligopoly. The most significant of these shifts merely transferred market control within the oligopoly. A brief survey follows. When Bayer purchased Monsanto in 2018 it dropped the company's 117-year-old highly tarnished name. Local operations have started to wear the Bayer brand. The merger of Dow and DuPont created Corteva Agriscience, which has started to put its brand on some facilities. According to Corteva, all Hawai'i research and development activities are being consolidated to the west side of Kaua'i, and 280 workers around the rest of the state (concentrated on Moloka'i) were laid off. Before the purchase of Syngenta by ChemChina was approved, Syngenta's local operations were taken over by Hartung Brothers, Inc. Hartung, an Iowa based company, acts as a contractor for Syngenta's Hawai'i-based operations. The sale to Hartung prior to completion of the ChemChina deal was triggered by the Committee on Foreign Investment in the United States' decision that Hawai'i locations must be exempt from Syngenta's purchase by ChemChina. Syngenta operations are bordering the world's largest missile range, purportedly sparking US military concerns about proximity to a Chinese state-owned company. Finally, in 2016 BASF sold their Kaua'i operations to Beck's, the fourth largest seed company in the US. Beck's has collaboration agreements for research and distribution with the larger agrochemical-seed-biotech companies. Beck's has said they intend to continue the research BASF was doing and that there will be few changes in its operations.

Although companies that do contract work for the Big Four have in some instances taken over Hawai'i operations, this has not fundamentally altered the nature of operations or who they benefit. Much marketing emphasis since 2017 has been on supposed changes to the industry locally—for example, Hartung emphasizes that they are now "family owned." This appears more an exercise in public relations than a reorientation in companies' primary focus on agrochemical-seed-biotech operations and primary generation of profits for the Big Four oligopoly.

As of 2015, the agrochemical-seed-biotech industry occupied at least 23,728 acres across Hawai'i. There has not been a comprehensive assessment of agricultural lands since 2015, and even state entities that lease lands to the industry do not keep accurate inventories; thus, these figures have likely declined somewhat since 2017 mergers. As of 2015, the industry's land footprint included around 13,230 acres on the island of Kaua'i, 7,350 acres on O'ahu, 2,340 acres on Moloka'i, and 750 acres on Maui. On the island of Kaua'i, which has around 21,310 total acres in crop production, the seed industry comprises 62 percent of agricultural land use, not including pasture land. Prior to Dow and DuPont's consolidation, this figure stood at 65 percent for the small island of Moloka'i (Melrose, Perroy, and Cares 2016). Some local agricultural businesses—such as a shrimp farm on the island of Kaua'i—sublease hundreds of acres of "public lands" (those confiscated in the US-backed overthrow of the Hawaiian Kingdom) to seed companies. Local farmers also have contracts to grow crops for the industry. Thus, the amount of land being occupied for genetically engineered corn seed production may not be accurately reflected by industry leases and ownership. Moreover, the industry subleases lands they control back to small farmers, a notable strategy that has ramped up since 2013 in response to criticism of their landholdings.

According to the industry, they plant on only about 25 percent of arable lands under their control at a time, which amounts to around 10 percent of all leased acres (including arable and nonarable lands). The seed industry reported a high of 6,910 harvested acres in 2011–2012, down to 2,565 expected harvested acres in 2019–2020 (USDA National Agricultural Statistics Service 2020). Though this figure does not account for all of the area they occupy or for their experimental field trials, it is a marked drop in production. In addition to industry consolidation and global market trends, the decline in harvested acreage is at least in part due to social movement pressure to recede from fields that are most proximate to residential areas.

In step with the rapid commercialization of genetically engineered crops across the United States, the estimated total value of Hawai'i's seed industry grew at an average annual rate of 18.5 percent from 2000 to 2013 (Loudat and Kasturi 2013). The industry's estimated value peaked in 2011–2012 at $241.6 million. In 2019–2020 the estimated value of the seed industry had dropped to $110 million (USDA National Agricultural Statistics Service 2020). The years prior also witnessed significant declines in the estimated value of the industry. Again, while this is partly reflective of global-local consolidation, some of the downturn is undoubtedly tied to growing resistance to the industry in the islands. Moreover, these value rates are fuzzy figures, estimated by the USDA based on company reported data. There is no consistent market measurement of value rates because outshipments of seeds from Hawai'i are not sold as regular commodities. Instead, the USDA does an annual survey of companies, asking them about their operation budgets and crop acreage and the value of the seed they shipped. Based on this company-reported data, they generate estimates of industry value. Because these are estimates of economic activity versus actual production, they do not necessarily correlate directly to how much companies are producing and selling. For example, in 2009–2012 the industry was making infrastructure investments that boosted their purported value. Other agricultural industries are not reported to have a higher value when their budgets are higher or more is being spent on infrastructure.

The industry also self-reports how many and what types of jobs it generates. According to a 2013 report funded by the trade group Hawaii Crop Improvement Association, the industry employed nearly 1,400 people, with 43 percent of jobs being part-time (Loudat and Kasturi 2009, 2013). In 2019, the industry reported employing 1,150 part and full-time workers, a decline attributed to consolidation (Yerton 2019). This equates to around 9 percent of the state's recorded 12,500 farmworkers and 0.17 percent of all jobs. While a small figure, these jobs are often concentrated in communities that lack other employment options. For instance, on the island of Moloka'i, Bayer (Monsanto) and Corteva (DowDuPont) were the largest formal employers, providing an estimated 11 percent of jobs (though subsistence activities are arguably more significant to livelihoods and the economy on Moloka'i) (Johnson 2014). With the merger of Dow and DuPont, many of these jobs on Moloka'i were terminated. While occupying far less land and employing far fewer people than the plantations of old, the agrochemical companies have stepped into the role of the reigning capitalist agribusiness in the islands.

Impacts to Human Health

Like capitalist plantations prior and other large monocrop agribusiness in the islands today, GMO seed development and field trials are associated with intensive pesticide use. Evaluating and commercializing new genetically engineered corn lines requires several generations of increase to get the necessary quantities of seed. Hawai'i's year-round growing season allows companies to speed up this process, but because there is no winter setback in pest populations, companies use large amounts of pesticides to protect their lucrative seeds. James Brewbaker explains the high application of pesticides: "Unique pesticide regimes are imposed by the seed industry, reflecting the value and the high quality expected of germplasm produced by the seed industry. Seeds are treated before planting, and preemergence insecticide is usually incorporated at the time of planting for control of leafhoppers and thrips. Subsequent insecticide and fungicide treatments are applied on a 5–7 day regime, and scouting for any new outbreaks is rigorous throughout the growing season. One result of this insecticide regime is that many predators and parasites of corn pests are also eliminated or reduced in population" (Brewbaker 2003, 69).

The inbred varieties that are grown for breeding purposes in Hawai'i are also more vulnerable to pests and disease. Moreover, with herbicide resistance being the vast majority of tested traits, experimenting with combinations of agrochemicals on crops is central to field trials. Multiple permits were issued in Hawai'i for testing of Dow's 2,4-D crops and Monsanto's dicamba-resistant crops (Freese, Lukens, and Anjomshoaa 2015). Though much of the acreage that the agrochemical companies occupy lies fallow, fields are typically still sprayed with pesticides to keep insect populations down and weeds from growing. According to University of Hawai'i tropical plant and soil scientist Hector Valenzuela, "To grow either seed crops or test crops, you need soil that's essentially sterile" (Pala 2015).

While the public does not have access to precise data about agrochemical industry pesticide use, it is known that large amounts of both restricted-use pesticides (RUPs) and general-use pesticides are used by the industry. RUPs are those deemed by the EPA to need a special permit and protective equipment for application due to their known harmful impacts on human health and the environment (40 CFR §152.170). According to voluntary reporting by Syngenta, DuPont, Dow, and BASF, they used 36,240 pounds of restricted-use pesticides on the island of Kaua'i alone in the twenty-month period from December 2013 to July 2015. One year of voluntary RUP reporting included

S-metolachlor (3,895 pounds), atrazine (1,876 pounds), chlorpyrifos (1,686 pounds), paraquat-dichloride (475 pounds), alachlor (466 pounds), methomyl (320 pounds), mesotrione (206 pounds), permethrin (183 pounds), lambda-cyhalothrin (113 pounds), chlorantraniliprole (85 pounds), zeta-cypermethrin (28.3 pounds), esfenvalerate (28 pounds), tefluthrin (25 pounds), S-cyano methyl (22 pounds), beta-cyfluthrin (13 pounds), and cyfluthrin (2 pounds) (Kaua'i Good Neighbor Program, n.d.). Companies reported planting on 1,841 acres. This equates to a total 9,423 pounds over 1,841 acres, or 5.12 pounds of active ingredient RUP per acre. This does not include surfactants and adjuvants, which can also be harmful to human health and the environment. It also does not include general-use pesticides. The spread of this use likely varies immensely, with higher usage on some acres and lower use on others. Looking at the data for each company individually, approximate higher amounts were used by Syngenta (4,795 pounds / (0.1 × 3,000 acres) = 15.98 pounds per acre), followed by Dow (3,157 / (0.1 × 4,500 acres) = 7 pounds per acre), DuPont Pioneer (1,252 pounds / (0.1 × 4,000 acres) = 3.13 pounds per acre), and BASF (218 pounds / (0.1 × 1,000 acres) = 2.18 pounds per acre. These estimates are based on the approximate rate of cropping 10 percent of leased acres, as publicly stated by companies. Based on this company-reported data, the Hawai'i Center for Food Safety estimates that Kaua'i's seed corn fields receive seventeen times more restricted-use insecticides than corn grown in the continental United States (Freese, Lukens, and Anjomshoaa 2015). Other comparative estimates are lower; all are imprecise, given incomplete data, especially with respect to acreage use (JFFG Draft 2016, 26). Corresponding to the very recent decline in land use and relocating of some parent seed production, the amounts of pesticides used have likely also declined significantly just in the past couple of years.

At least fifteen of the twenty-two restricted-use pesticides known to be used by the industry have been linked to cancer by the American Cancer Society and American Academy of Pediatrics (Skolnick 2013). Of the seven most commonly used RUPs, the EPA suspects six of being endocrine disruptors. Between them, the seven have been linked to "neurological and brain problems and damage to the lungs, heart, kidneys, adrenal glands, central nervous system, muscles, spleen and liver" (Koberstein 2014). Even at very low exposure levels, the three most widely used RUPs have all been strongly linked to serious health impacts.

The volume of general-use pesticides (GUPs) used by agrochemical companies is unknown to the public. General-use pesticides are those that can be purchased and used without a permit, including 2,4-D, neonicotinoids, dicamba, acetochlor, and glyphosate. Though less regulated, GUPs have proven serious health and environmental impacts. For example, neonicotinoids are highly toxic to bees, and other GUPs are carcinogens and endocrine disrupters. The Center for Food Safety estimates that total pesticide use is likely four times higher than reported RUP use (Freese, Lukens, and Anjomshoaa 2015). From a class action lawsuit on Kaua'i, it was divulged that when general-use pesticides are accounted for, at least ninety pesticide formulations with sixty-three different active ingredients were used by DuPont (now Corteva) between 2007 and 2012. Pesticides were applied 250–300 days per year, at an average of ten to sixteen applications per day (Jervis and Smith 2013).

Companies have also used illegal pesticides, potentially more frequently than actually known. Allegations have become so frequent that the EPA was recently compelled to pursue a multifaceted inquiry into several companies. A first result of the inquiry has been Monsanto pleading guilty to knowingly storing and using a banned pesticide, methyl parathion (Penncap-M), on the island of Maui in 2014. Monsanto ordered workers to enter fields just seven days after Penncap-M was used ("Monsanto to Plead Guilty" 2019). Prior to being banned, the federally mandated waiting time to enter a field after use of methyl parathion was thirty-one days. Methyl parathion is highly toxic by inhalation.

Of particular concern is the "stacking" of pesticides, or their mixing into cocktails with more harmful and almost entirely untested and unregulated synergistic effects (Vallianatos 2014). A State of Hawaii–sponsored fact finding group found that "two or more pesticides are sometimes combined in a single spraying solution . . . The exact mixtures used on Kaua'i are not known, so no cumulative risk assessments regarding them can be referred to at this time" (JFFG 2016, 25). Exposure to combinations of an "ever-changing kaleidoscope of pesticides," as described by ecologist Sandra Steingraber, may be most harmful to human health (2016). For example, a recent review of the scientific literature indicated that the cumulative effects of individual noncarcinogenic chemicals "acting on different pathways, and a variety of related systems, organs, tissues and cells could plausibly conspire to produce carcinogenic synergies" (Goodson et al. 2015, S258). Rather than a science-based approach to evaluating harm, the EPA determines which and how pesticides can be used based on tests of singular active ingredients, not tests of actual mixtures being used (see chapter 6).

The pesticide-intensive operations of the agrochemical-seed-biotech industry are in large part located upwind and adjacent to residential areas, including schools, homes, and hospitals, as well as shorelines and waterways. According to Hector Valenzuela, "The mosaic style of planting, with small plots of corn surrounded by expansive bare fields, means that the risks and actual levels of pesticide drift, runoff, and dust pollution are considerably greater than the levels observed under normal commercial farms in Hawai'i" (2016, 6).

It is common practice for agrochemical-seed-biotech companies to plow over entire fields, leaving them barren and susceptible to generating pesticide-laden dust in Hawai'i's high wind conditions. A 2011 class action lawsuit brought by over 150 residents on the west side of Kauai alleges that pesticide-laden "excessive fugitive dust" from DuPont Pioneer (now Corteva) had blown into their homes on an almost daily basis for more than ten years (Jervis and Smith 2013, 5). Much of the land used by agrochemical-seed-biotech operations has not recovered from sugar and pineapple plantation days, is without healthy topsoil, and is still contaminated by older pesticides.

Numerous local doctors have submitted official statements expressing concern that they may be witnessing impacts of pesticide exposure in communities living near fields. Impacts potentially include higher than average rates of rare birth defects, miscarriages, unusual cancers, respiratory and hormonal problems, and recurring dermatitis and nosebleeds (public testimony on Kaua'i Bill 2491; Pala 2015). The following are excerpts from emails and testimonies submitted by practicing doctors from the west side of Kaua'i during debates over Kaua'i County's pesticide disclosure Bill 2491.

> Some of our individual and collective concerns: We have observed a higher than normal occurrence of birth defects and miscarriages. High rates of very severe gout in healthy populations. Rare types of cancers in a higher than expected incidence for our small population of patients. Almost daily reports of respiratory symptoms in patients that have no history of these respiratory illnesses. Many of these respiratory symptoms are not responding to the pharmacological interventions typically prescribed. Hormonal changes including excessive facial and body hair on women and higher levels of infertility. Reoccurring nose bleeds in children. Reports of a "metallic taste" in their mouths and recurring dermatitis. (Kapono Chong-Hansen MD, Carla Nelson MC, Nernard Riola MD, Rick Goding MD, Surcchat Chatkupt MD, Margie Maupin NP)

> Of particular concern is the incidence of serious cardiac malformations; particularly those that result from early embryogenesis defects that have occurred in our population the last three years. We have had 5 major cardiac defects that have required early extensive surgical repair in San Diego the last 3 years: 2 cases of Transposition of the Great Vessels, 1 Hypoplastic left heart, 1 Hypoplastic Right heart with heterotaxy and 1 severe pulmonary stenosis. The transposition cases and hypoplasia cases are considered defects that occur in early first trimester. While cardiac

> birth defects are the most common birth defects these particular types of lesions are rare. Recent CDC statistics puts transposition at 1/3,300 births, hypoplastic left heart at 1/4,344 births and hypoplastic right at 1/17,000 births. In the last 3 years we have had about 750 deliveries; this gives us an incidence of 53/10,000 births for these 4 defects. National U.S. data shows an incidence of 5.5/10,000 births, *so we have 10 times the national rate*." (Emphasis added; names not included to protect privacy; original email can be read at http://garyhooser.com/birthdefects.html)

> I have been told in no uncertain terms by the head of the Tumor Board at Wilcox that there is a significantly higher incidence of leukemia and multiple blood disorders that are considered "cancers" by some hematologists, but not tracked by the registry on the West side. I have also been informed by a practicing oncologist that the same situation exists for incidents of lymphoma on the island. The rate of serious birth defects on the West side is indisputably higher (10x higher in the case of some serious cardiac defects requiring surgery) than the national average. These have trended significantly higher in the last 5 years. Again, these data are only a starting point for a true study. They require significant investigation to help uncover the causes, and a longer timeline to establish a trend. (Email submitted to the *Garden Island* newspaper by a practicing MD)

Medical professionals on the west side of Kaua'i have repeatedly emphasized that, while concerned, they are not certain about impacts and that data and local study are severely lacking. There is simply not enough information for definitive determinations, a conclusion that was also reached by the State of Hawaii–sponsored Joint Fact Finding Group, which spent a year assessing the situation. There have been no local epidemiological studies. There is no mandatory disclosure of when, where, and what types of pesticides are used, making the extent of resident exposure unknown and significantly impeding health and environmental study.

Communities neighboring agrochemical industry operations face what Sandra Steingraber calls the "miasma of uncertainty" (1997, 71). Not only is data severely lacking, but it can be nearly impossible to definitely prove direct environmental causes and effects. Some amount of scientific uncertainty is a fact. This is well understood by polluting industries, who have a long history of manipulating doubt in cases ranging from DDT and lead poisoning to tobacco and climate change (Oreskes and Conway 2010; Markowitz and Rosner 2003). The ability to manipulate uncertainty has provided a refuge to countless corporate criminals and served as a barrier to enacting robust environmental health protections (Seager 2003, 964). It is those who are on the frontlines of environmental destruction and pollution that suffer the long consequences of proving harm.

The State of Hawaii itself has at times appeared to act as a buffer for the industry on questions of exposure and harm. Rather than err on the side of scientific investigation and protection of human health, it has more frequently championed the claims and interests of the agrochemical conglomerates. For example, in the midst of controversy the Hawaii State Department of Health (DOH) announced conclusively that islandwide Kaua'i cancer rates are no higher than the rest of the state. The claim has been cited repeatedly to assure communities that there is no harm being caused by agrochemical-seed-biotech operations. Glaringly, the report did not look at a breakdown by census track—in other words, it did not look specifically at communities living near fields. Several local doctors expressed discontent over the state's assertion: "[T]he DOH statement is not definitive and does not resolve this issue in the minds of the physicians on this island" (west side doctor, email communication, 2013). One local MD wrote in an unpublished letter to the newspaper: "there is a serious disconnect between the clinical observations of our island physicians and the Tumor Registry numbers reported . . . The blanket press release from DOH, seems to end the questions, when in fact, the questions need more consideration." Concerns of west side physicians over high rates of rare birth defects also remain unreconciled with official state records. Different divisions of the Hawaii State Department of Health have supplied conflicting data to the public, and no birth defect data was kept from 2005 to 2010 due to budget cutbacks (JFFG Draft 2016).

Several acute pesticide exposure events have also been of grave concern, and also with dubious response from the State of Hawaii. On the west side of Kaua'i, on at least three occasions in 2006 and 2008 children at Waimea Canyon Middle School were taken to the hospital after smelling noxious odors from a neighboring Syngenta field. In public testimony, a teacher at the school recalled:

> In November of 2006, after the neighboring fields had been sprayed, sixty-one students were sent to the health room complaining of headache, nausea, vomiting, diarrhea, and dizziness. Thirty seven were sent home. The spraying continued, as did the illness spikes throughout the year. On January 25, 2008 while playing on the P.E. field, my 2nd period class and I noticed a strong odor. Many of the students complained that it smelled like "garden fertilizer." Within minutes of smelling it, my class and I,

> along with a nearby custodian, started to feel dizzy and ill. At that point I reported the incident to our school administration. Over the course of the day, approximately 60 students were sent to the health room. Twelve people, including a staff member, were sent to the emergency room and countless others suffered silently. Later that day I found out that every student but one in my 2nd period class had become very ill and some had to go to the emergency room to be treated for nausea, burning eyes, difficulty breathing, vomiting, headache, diarrhea and extreme dizziness . . . I felt ill for the next four days. (Public Testimony 2008)

Numerous incidents related to pesticide spraying by Syngenta were reported to the Department of Agriculture at the school from 2006 to 2008. In a severely delayed report, the Hawai'i Department of Agriculture suggested that causes were inconclusive and focused investigation most heavily on the commonly occurring plant stinkweed (*Cleome gynandra*) (Li, Wang, and Boesh 2013). Stinkweed odors have long been present and have never been known to cause the severe symptoms suffered by students and teachers. A former senior health and science adviser to the EPA responded that there is "no quantitative or literature evidence" to support the theory that stinkweed was a culprit in the school poisonings. The former EPA adviser called it an "unusual emphasis" (JFFG Draft 2016, Appendix 2, 146). He concluded that "while there is no definitive cause for the health symptoms reported in Waimea, they were far more likely related to pesticide exposures than from exposure to stinkweed organics, other plants or their decomposition products" (JFFG Draft 2016, Appendix 2, 147). Following protests by teachers and parents, negative press, and the first ever temporary restraining order filed by the Hawaii State Teachers Association, Syngenta ceased planting the adjacent field and nothing similar has occurred since (Gregg 2008; Perry 2013).

While the state's study of Waimea Canyon Middle School events was egregiously delayed, its air sampling has still consistently detected chlorpyrifos and other pesticides used by agrochemical-seed-biotech companies (Li, Wang, and Boesh 2013). Though detected at legally permissible levels, recent studies suggest that low-level exposure to organophosphate insecticides like chlorpyrifos can have profoundly negative impacts on children's neurological development, especially *in utero* (Eskenazi et al. 2007; Bouchard, Bellinger, et al. 2010; Bouchard, Chevrier, et al. 2011; Engel et al. 2011; Roberts et al. 2012). A major long-term study by University of California scientists found a 60 percent increased risk of autism spectrum disorder in children of mothers who lived within a mile of fields sprayed with organophosphates during their pregnancies (Shelton et al. 2014). Chlorpyrifos is known to volatilize and drift for days or months after application, exacerbated by Hawai'i's warm climate and windy conditions. Given the dangers of chronic low-level exposure, its ability to drift, and the large amounts of chlorpyrifos used by agrochemical-seed-biotech companies, many believe it to be one of the greatest threats to human health of the industry's operations. Single incidents of acute high-dose exposure to pesticides like chlorpyrifos can also have long-term health impacts. In one known such case, at least ten contract migrant field workers were sent to the hospital after being exposed to chlorpyrifos at a Syngenta operation (JFFG Draft 2016, 86). There is no public knowledge about their recovery or current whereabouts.

While direct impacts to human health from agrochemical-seed-biotech operations have not and cannot be comprehensively and scientifically studied, due most especially to industry secrecy and state complicity, there is no question that the pesticides used in large amounts are highly dangerous to people and the wider environment.

Environmental Impacts

Pesticides significantly impact soil health, marine and freshwater ecosystems, bees, biodiversity, endangered and other species. Again, no comprehensive studies have been conducted on local environmental impacts of agrochemical-seed-biotech operations. Hawai'i's unique ecology and geography present particular concerns. The islands' steepness, intense rainfall, flashy streams, and small drainage basins all contribute to rapid dispersal of contaminants on land into waterways (Oki 2003; DOH 2016). Many of the pesticides used by agrochemical operations are known to persist in the environment and to contaminate water sources. A 2014 Department of Health stream sampling study tested four locations on drainage canals downstream of agrochemical operations, concluding that "detections at these sites were compared to reported restricted use pesticide (RUP) application under Kaua'i's Good Neighbor program. Five restricted use pesticides were detected at one or more of these sites, and three, atrazine, metolachlor and chlorpyrifos were reported to have been used by seed crop operators a few weeks prior to sampling" (DOH 2014, 21–22).

Local environmental scientist Carl Berg notes that "since many of the pesticides are known hormone disrupters, the effects on larval and juvenile development of fish could have serious implications to Hawaiian stream species and nearshore fisheries" (2016, 2). Mudslides along coastal agricultural fields due to unpermitted soil grubbing by companies have also impacted aquatic ecosystems and commercial and subsistence fishing (Van Voorhis 2011). A teacher at

Waimea Canyon Middle School reports documenting impacts of pesticides on his classes' aquaculture project:

> Although from healthy stock, fish that we were raising began showing signs of genetic deformity, stunted size, and rarely grew to adulthood without dying. The adult breeders were kept in an 800 gallon tank and did not seem affected. The fry were all in 100 gallon tanks. All tanks had the same filtration system and water. These deformities did not occur during the period of time that the ag lands were planted with sugar cane. They only began to occur after the planting of GMO corn and application of organophosphate pesticides that is done weekly (sometimes twice weekly). A complaint was filed with the Dept. of Ag and the extent of their investigation was to take pictures of floating dead fish. The project was put on hiatus and the adult breeders moved to another location. Since this move multiple generations of fry have been raised to adulthood without one instance of genetic deformity or die off. (Public Testimony 2008)

Dozens of pesticides used by agrochemical companies are known to be toxic to bees and pollinators (Valenzuela 2016). In 2000, a sudden bee die-off on the west side of Kaua'i was attributed in an industry consultant's report to insecticide use by agrochemical companies (JFFG Draft 2016, 49). In 2015, a high school student's science project tested the amount of glyphosate in honey from domesticated and wild beehives around the island, concluding that hives on the west side were more frequently contaminated. An official organic farm certifier reports having to deny organic certification to beekeepers due to pesticide contamination from surrounding agrochemical company fields. There are anecdotal reports of harm to other species as well, such as rare occurrences of disease observed by hunters and fishers. According to the Joint Fact Finding Group's report, "Several local residents on the Westside have reported what they believe may be an unusual number of dead or sick owls but no samples of blood or tissues for pesticide residues appear to have been taken to date" (JFFG Draft 2016, 48).

In addition to the impacts of pesticides, GMO operations can spread engineered traits to relative species through gene flow. The Hawaiian Islands are considered both a biodiversity hotspot and the "endangered species capital of the world" (Scheuer and Clark 2001). Close to ten thousand species are endemic, found nowhere else on the planet, and 437 species are listed as either threatened or endangered (Evenhuis and Miller 2015; Freese, Lukens, and Anjomshoaa 2015). In one example of gene flow in Hawai'i, the EPA has recognized that Bt cotton has the potential to crossbreed with a wild endemic variety of cotton (Gibson 2014). Cross-pollination with other agricultural crops is likewise well documented. Of twenty thousand papaya trees tested by a citizens' group on Hawai'i Island, 50 percent were shown to be genetically engineered. Eighty percent of trees tested were from organic farms that had not intended to cultivate genetically engineered varieties (Barnett-Rose 2015). In fact, one of the very reasons Hawai'i is considered a favorable location for open air testing of genetically engineered crops is that, in contrast to places like the Midwest, it is isolated from commodity crops that could be easily contaminated.

Spread of genetic traits to wild and cultivated relatives is of particular concern in regard to outdoor field testing of crops engineered to produce pharmaceuticals. A 2006 lawsuit brought by Earthjustice against the USDA revealed that companies were conducting open-air field trials of corn and sugarcane containing hormones, HIV and other vaccines, cancer-fighting agents, and other proteins to treat human illness (*CFS v. Johanns* 2006). There have been occasions of such pharmaceutical crops escaping field trials and being grown amid commercial crops intended for human consumption (Achitoff, Kimbrell, and Wu 2015). While biopharmaceutical testing is no longer being conducted in Hawai'i, lack of public disclosure leaves residents uneasy and speculating about unknown risks.

Given the significant environmental and human health consequences of being GMO Ground Zero, it must be asked why Hawai'i is at the epicenter of the agrochemical-seed-biotech oligopoly's global chains of production. It is often stated by critics and proponents alike that the industry operates in the islands merely because of a favorable climate. While Hawai'i's year-round growing season clearly facilitates rapid development of herbicide-resistant seeds, this narrow explanation insinuates that the situation is just natural and invisibilizes much more essential social and historical circumstances. The next three chapters show how the conditions for the agrochemical industry's occupation of Hawai'i are not reducible to sunshine and natural competitive advantage, but premised upon exploitative, oppressive, and imperial social relations. There is also nothing inevitable about Hawai'i continuing on this trajectory.

Fig. 1. Agrochemical/GMO industry research and seed development on the westside of Kaua'i, where the industry occupies thousands of acres. (Photo by and reproduced with permission of Samuel Shaw, 2013)

Fig. 2. "No trespassing" sign along the fence around public lands managed by the State of Hawaii's Agribusiness Development Corporation and leased to agrochemical companies. (Photo by and reproduced with permission of Dylan Hooser, 2013)

Fig. 3. Agrochemical/GMO industry fields on the westside of Kaua'i. It is common practice for companies to plant in small plots surrounded by expansive bare fields, resulting in excessive dust. (Photo by and reproduced with permission of Samuel Shaw, 2013)

Fig. 4. DuPont Pioneer (Corteva) fields (left side of image) upwind of the community of Waimea on the west side of Kaua'i. In 2011, more than 150 Waimea residents filed a class action lawsuit against DuPont, alleging that pesticide-laden dust has blown into their homes on an almost daily basis for more than a decade. (Photo by and reproduced with permission of Samuel Shaw, 2013)

Fig. 5. The state's largest coffee plantation, which leases land to DuPont/Corteva on coastal portions of their property on the southwest side of Kauaʻi (upper part of image). Kauaʻi Aggregates Quarry is also pictured. (Photo by and reproduced with permission of Samuel Shaw, 2013)

Fig. 6. Agrochemical/GMO industry fields on the west side of Kauaʻi, surrounding the community of Kekaha. Environmental injustices in Kekaha are numerous and compounding—the Pacific Missile Range Facility and the island's only landfill are also located in the community. (Photo by and reproduced with permission of Samuel Shaw, 2013)

Chapter 4

Imperialism and the Making of a Plantation Economy

The story of how and why Hawai'i has come to be an epicenter of agrochemical-seed-biotech production begins with the dual processes of imperialism and capitalism. Attending to beginnings of the imposition of capitalism and empire is necessary for understanding the islands' present and the possibility of truly different futures. As Samir Amin writes, for both the "West" and the "Rest," capitalism was not a foregone conclusion simply waiting to emerge (Amin 2009). Just as capitalism's development in the English countryside required the massive dispossession of peasants from the commons (Wood 2002; Marx 1992), so too did its imperial spread involve violent expropriation. Knowing the recent history of imperialism and capitalism, and their ongoing contestation, denaturalizes the current social order. It reminds us that much different kinds of social orders have existed in our human past, survive in our present, and are possible in our future.

What follows is not a comprehensive history of imperialism, capitalism, or their resistance in Hawai'i. This chapter is more limited and borrows from the work of others to review critical moments of imperial-capitalist compulsions and ideology bearing down upon the islands, including strategic accommodation, appropriation, and resistance by Hawaiians. It traces the emergence of an oligopolistic plantation economy—inseparable from US occupation of the islands—that persists to this day. An ōlelo no'eau, or Hawaiian proverb, instructs that the present is informed by the past—"I ka wā ma mua, ka wā ma hope: the future is in the past." Lilikala Kame'eleihiwa explains: "It is as if the Hawaiian stands firmly in the present, with his back to the future, and his eyes fixed upon the past, seeking historical answers for present-day dilemmas" (1992, 22–23). In order to know where we might go, we need to perceive where we have been.

Contact with Euro-American Imperialists

Indigenous Hawaiian society was a dynamic and evolving one, with multiple migrations from different places in the Pacific. Changes in class structure, governance, land and production systems, culture, and spiritual belief that took place over time were significant; "ancient Hawai'i" cannot be interpreted as a place stuck in time. While Hawaiian society was a product of numerous, sometimes ongoing Pacific migrations, for over a millennium the islands' peoples lived in steady balance with the rest of the web of life. Spirituality and sacredness structured peoples' relationships with one another and the 'āina. The word *'āina*, commonly translated as "land," in one more literal meaning is "that which feeds or nourishes." Carlos Andrade explains: "The dual aspects of spirit and mind remain inseparable from Native understandings of 'āina, which nourish Hawaiian identity, and mystically and genealogically connect the people to the islands and to generations of ancestors who came before them" (2008, 76). Oral history tells of Kānaka Maoli (Native Hawaiians) as genealogical descendants of the earth, sea, and sky. This relationship of interdependency, reciprocity, and stewardship structured indigenous systems of living. "The 'āina (land) is the eldest sibling, and therefore responsible for protecting and feeding the younger ones. As younger siblings, Hawaiian people inherit a kuleana (responsibility) to mālama (care for, keep, obey, pay heed to) 'āina and kalo (taro plant)" (Andrade 2008, 25).

Mālama 'āina (care for land) was fundamental to maintaining pono, or balance between maka'āinana ("the people living on the land" or commoners), ali'i (chiefs or ruling class), kahuna (priests and experts), deities, and 'āina. According to Hawaiian political and language scholar Noenoe Silva, different from its common English translation as "righteous," pono meant that all was in harmony and carried with it an understanding that such balance was dependent upon all being fed, taken care of, and healthy (Silva 2004). A high degree of mutual obligation and concern for both people and land was implicit in such worldview. Lilikala Kame'eleihiwa writes, "Because the 'Āina is both the Ali'i Nui and the maka'āinana, as well as the elder and the younger siblings, Mālama 'Āina, in traditional times, was truly to care for and serve one another" (1992, 32).

Structured by relationships of reciprocity, indigenous Hawaiian production was organized cooperatively around 'ohana, or extended family units. Commoners freely accessed land, water, sea and forests, with the commons deliberated among maka'āinana and konohiki (heads or managers of land divisions) (Andrade 2008). Maka'āinana were able to move between ahupua'a (land units), and extensive kinship networks allowed them to relocate if they were unprosperous or at odds with konohiki. If famine or other imbalances arose, ali'i would be held accountable and deposed (Kame'eleihiwa 1992). Evolving indigenous Hawaiian society was not free from class hierarchy or conflict. However, it was defined by beliefs and structures of holism and collectivity, self-determination, and reciprocity. Its sophisticated systems of production and distribution were designed to ensure that all had enough and that careful stewardship and sacred reverence for the earth was maintained. It was a society in which the logics of capitalism—of unabated exploitation of land and people for personal gain, extreme individualism, absolute private ownership, accumulation of wealth for wealth's sake, and the deprivation of many alongside excess riches for very few—would have been structurally impossible and culturally unintelligible.

In the late eighteenth century Hawaiian society was undergoing internal change and instability in the warring, conquering, and eventual uniting of all the islands under Mō'ī (paramount chief, or translated by Beamer as "a succession of the supreme") Kamehameha I. At the same historic moment that Kamehameha was consolidating rule and governance structures were being reorganized, Hawaiians were also negotiating the increasing presence of imperial Euro-American powers. According to Hawaiian oral history, white foreigners were traveling to Hawai'i up to thirty generations before Captain James Cook arrived in 1778 (Beamer 2014). Moreover, there is evidence that prior to Cook's arrival, expert Hawaiian navigators were already exploring vast territories and engaging with other cultures. As Hawaiian studies scholar Kamanamaikalani Beamer writes, Kānaka Maoli may have "discovered" the haole (whites, foreigners) before they were "discovered" (Beamer 2014, loc. 435 of 5334). While earlier contact is likely, in the late eighteenth century Euro-American imperialism and capitalist expansion was heightening in the Pacific. Kānaka Maoli showed remarkable skill at navigating new outside pressures from the very earliest incursions of Euro-Americans in the islands. When violence between European ship captains and ali'i started to become a problem, Kamehameha captured foreigners and strategically gained their allegiance as trusted lifelong advisers in attaining knowledge about Western cultures and systems (Beamer 2014). This was to set a lasting pattern of ali'i intelligently learning about and negotiating the political-economic structures of Euro-American imperialists in order to best maintain Hawaiian independence, land, culture, and power.

Hawaiian power in and over the islands remained strong in the first decades of increasing contact, even as they navigated widespread death resulting from disease brought by Euro-Americans. Prior to 1778, Hawaiian society sustained a dense population—somewhere between 400,000 and 800,000, perhaps even close to where it stands today (Beamer 2014). Within the first fifty years of increasing contact with white foreigners, the native population declined 70 percent by conservative estimates. By the end of the nineteenth century, only 5–10 percent of the pre-Cook Hawaiian population remained (Swanson 2014). Depopulation and the resulting collapse of production systems shook the foundations of what had been an "intensely populated, independent, and sustainable society" (Beamer 2014, loc. 926 of 5334). As historian Samuel Kamakau vividly depicts: "To a people living happily in a pleasant land with purple mountains, sea-girt beaches, cool breezes, life long and natural, even to extreme old age, with the coming of strangers, there came contagious diseases which destroyed the native sons of the land . . . the land has become empty; the old villages lie silent in a tangle of bushes and vines, haunted by ghosts and horned owls, frequented by goats and bats" (1992, 416). Decisions by Kānaka in these decades of tumult and horror need to be understood entirely in this context of uncontrollable, tragic loss.

As pandemics struck Hawaiian society and the land was increasingly emptied of its people, global commerce and new populations were simultaneously establishing. Euro-American imperialists brought not only deathly disease, guns, and novel gadgets but also ideas and compulsions of capitalism. Hawai'i's introduction to global trade was as a provisioning station for fur traders. As the market in fur closed, traders turned their attention to extraction of the fragrant 'iliahi, or sandalwood. The resource extraction economy of sandalwood could be sustained only temporarily, but a booming Pacific whaling industry gave rise to populous port towns and lasting demand for goods and services. By the mid-1840s, nearly five hundred whaling vessels passed through Hawai'i's ports annually. The foreign population grew from under one hundred in 1823 to 1,500 by 1850, boosting demand for local products and increasing the circulation of money in the islands (MacLennan 2014). Many foreign traders were "growing rich rapidly" in the new Pacific capitalist frontier (Kent 1993, 24).

Debt was used early by imperial powers to extract wealth and assert commercial interests over Hawai'i. In 1826 and 1829 US warships arrived, demanding payment for debts that allegedly were owed by ali'i in the sandalwood trade. Anthropologist Marion Kelly describes the arrangement as fraudulent and as ensnaring the nation in a cycle of debt that further forced it open to imperialist-capitalist incursions: "The very earliest experiences of the Hawaiian Nation with the sandalwood trade reveal a direct relationship between foreign investment and local indebtedness. The value of the goods received by the Hawaiian chiefs had been paid for, perhaps several times over.

With sandalwood resources exhausted, recovery from debt within any foreseeable future was impossible" (1994, 16). The first treaty between the Hawaiian Kingdom and the United States included a list of debt claims and demand of payment, backed by the threat of violent enforcement—a harbinger of things to come.

Imperial-Capitalist Enclosures, Adaptations, and Resistance

The nineteenth century was one of competitive Euro-American imperialism throughout the Pacific—an "orgy of national enslavement" as Tom Coffman put it (1998, 63). Frequent military incursions onto Hawai'i's shores by Britain, Russia, France, and the United States served as a reminder that independence was always tentative. Moreover, accused debt ensnared Hawai'i in the imperial-commercial economy even before it was recognized by colonial powers as a sovereign nation. Militarily imposed agreements for repayment of debt set in place "a cascade of commercial pressures" (MacLennan 2014, 58), forcing the mō'ī to manage the new requirements of generating capital. Concurrently, Hawaiian social structures and belief systems were disrupted by disease and death (Osorio 2002, 12). In this context of extreme upheaval and always looming colonial threat, Hawaiian leaders worked to maintain indigenous sovereignty for almost all of the nineteenth century. They selectively appropriated the tools of the colonizers "as a means to resist colonialism and to protect Native Hawaiian and national interests" (Beamer 2014, loc. 193 of 5334). Their strategic adaptation of Western law and property in some ways opened the way for and in other ways likely delayed the engulfing system of capitalism.

Hawaiian ali'i incorporated Euro-Americans into their closest circles in attempt to gain knowledge of and navigate around the systems and cultures of imperial powers. Missionaries were especially prominent advisers, and brought with them the bibles of both Christianity and capitalism. Christianity was not wholly embraced or supported; it was a matter of deep contention among ali'i and maka'āinana alike. Following the death of Kamehameha I, for instance, tensions over tradition and incorporation of new religion mixed with political and power struggles, leading to deadly battle. As a whole, Kānaka Maoli differed in their embrace or rejection of Christianity—some were skeptical while others were ardent converts (Beamer 2014). While much has been written about the influence and power of missionaries in Hawai'i, early missionaries were subservient to ali'i, recognizing that their projects relied entirely on ali'i support and that this support was tentative (Beamer 2014).

When ali'i turned to missionaries and other foreign advisers for counsel in

Western law and economy, they were given instruction in the purest of capitalist economic doctrine. Though it was ultimately not the allure of missionaries' and other colonists' ideas or ideals that compelled Kānaka accommodation of capitalist ideology, it is worth laying them bare. Missionary letters back to the mission and other Euro-American foreigners' writings tended to be extremely racist and ignorant of the sophisticated, sustainable, noncapitalist society that indigenous Hawaiʻi was. They frequently compared Hawaiian society to the backwardness of European feudalism. American naval officer Henry Wise chastised: "Without an incentive to greater efforts, the country languishes under the same species of feudal tyranny and extortion as in the days of their cannibal forefathers! The islands are rich and fertile; sugar, coffee, and tobacco flourish luxuriantly; and under any other system than the present, there could be no bounds placed upon the advantages and wealth that would follow" (1850, 336). Liberating commoners from "despotic lords" was viewed as a project of unshackling innate human drive to competitive and possessive individualism (Lee 1850, 32). Only these drives could compel man to labor in such a way that would reverse population decline and progress society: "Idleness, poverty and destitution of the means of advantageous labor . . . shortens the lives of the people" wrote missionary Artemas Bishop (Bishop 1838, 55). Of the communally based Hawaiian land tenure system, Bishop bemoaned, "what inducement can there be for the common people to make any effort to arise and shake off their degradation and poverty?" (Bishop 1838, 57).

Missionaries and other Euro-Americans singularly saw the compulsion of capitalism, private maximization of wealth accumulation, as the savior of a rapidly depopulating nation. When missionary William Richards was selected by King Kauikeaouli (Kamehameha III) to educate him and other aliʻi on the foreign world, he formulated a series of lectures based on the works of political economists such as Francis Wayland:

> God has designed men to labor, yet he has not designed them to labor without reward . . . as it is unnatural to labor without receiving benefit from labor, men will not labor continuously nor productively, unless they receive such benefit. . . . In order that every man may enjoy . . . the advantages of his labor, it is necessary . . . That he be allowed to gain all that he can; and 2. That, having gained all he can, he be allowed to use it as he will . . . This will require 1. That property be divided. (Wayland 1837, 108–109)

Wayland's writings not only preached the superiority of private property, but that free access to commons led to indolence and a decaying society: "The forest of an Indian tribe is held in common, and a few hundred families barely subsisted upon a territory which, were it divided and tilled, would support a million of civilized men. His bow and arrows, his wigwam, and his clothing are acknowledged to be, in the fullest sense, his own. Were these to be held, like his land, in common, the whole race would very soon perish, from want of the necessaries of life" (Wayland 1837, 109–110).

In the context of extreme depopulation and collapsing production systems, missionaries advised that only capitalist, profit-driven farming would increase productivity and advance society. Missionary J. S. Emerson lectured: "I think two things are requisite to make the people industrious and provident. First, the feeling that the land they cultivate is their own, for themselves and their prosperity. Second, the feeling that the land is of real value, and capable of being improved in value, and that all improvements are private gain" (in Osorio 2002, 32). The logic and language of "improvement" came directly from England's enclosure movement. Philosophers like John Locke argued that English peasant evictions and expulsions from common lands, as well as the commercialization of agriculture, were necessary to the process of turning "waste" into "improved" lands (Ross 1998; Wood 2002, 2012). To "improve" literally meant to do something for monetary profit. Capitalist landlord "improvers" were said to be the ultimate source of wealth, adding more to the "common stock" than they took away in the dispossession of peasant commoners (Wood 2002, 115). Centuries after Locke and the violent internal development of capitalism in England, European colonists' ideas of land improvement as rooted in private property and profit remained.

The racist, capitalist, individualistic ideas of Euro-Americans were not something Hawaiians passively accepted. As a recent wave of Native Hawaiian scholars[‡] especially have argued, rather than mere victims of imperialist-capitalist violence and ideology, Kānaka strategically incorporated "foreign ideas, political theories, and technologies" in order to protect Hawaiian national interests (Beamer 2014, loc. 219 of 5334). For instance, while largely accessed by ali'i, William Richards's political economy lectures were taken in selectively. They were a product of ali'i decisions and agency in understanding how countries with increasing influence over and ambitions in the islands were governed. Beamer asserts that this was "part of a larger plan to conduct politics on the international level so that Hawai'i would be respected by foreign nations" (2014, loc. 2118 of 5334). Telling of ali'i success in this regard, Richards

‡ See for example Beamer (2014), Beamer and Duarte (2009), Preza (2010), Beamer and Tong (2016), Banner (2005).

himself became highly loyal to the nation, traveling far abroad in attempt to secure Hawaiʻi's independence.

Aliʻi decisions to begin incorporating principles of Western law and property resulted from multiple converging factors. Pressures for changes in land tenure came early from European and American traders wanting ownership of house lots, wharves, and warehouses. These pressures were made impossible to ignore by the presence of foreign warships, including those demanding repayments of accused debts. On two occasions the British and US governments tried to secure fee simple land ownership by treaty, but King Kauikeaouli refused, explaining: "We indeed wish to give Foreigners lands the same as natives and so they were granted, but to the natives they are revertable and the foreigners would insist that they have them for ever" (in Banner 2005, 285). Many have argued that in the 1840s Hawaiian rulers had reason to believe that foreign takeover was imminent, and that they strategically sought to institute legal and property forms that would be recognized by colonial powers (Banner 2005; Silva 2004; Beamer 2014). Noenoe Silva writes, "Creating a nation in a form familiar to Europe and the United States was a necessary strategy of resistance to colonization because there was a chance that the nineteenth-century Mana Nui, or 'Great Powers' might recognize national sovereignty" (2004, 9). Law was a tool that allowed aliʻi to negotiate with and be recognized by imperial nations, as well as to protect their own authority and control the increasing number of foreigners residing in the islands. Moreover, aliʻi and mōʻī were concerned with reestablishing a state of pono, of harmony between social classes and with the land, by encouraging people to cultivate the land in the face of tremendous depopulation (Beamer 2014).

Though laws existed in pre-contact indigenous systems of governing, in 1839 the first formal body of written laws were issued by King Kauikeaouli. They were authored in the Hawaiian language, and aliʻi were actively involved in their creation. The laws defined the relationships between mōʻī, aliʻi, and makaʻāinana, and sought to protect commoners from abuses of power. They specified that land and resources were jointly owned by mōʻī, aliʻi, and makaʻāinana. The 1839 laws have widely been described as the first codification of private property rights based on Euro-American capitalist law. While partly true, the laws were not a mere replication of exclusive property law. Indigenous Hawaiian custom and practice regarding the commons and resource sharing was also enshrined. For instance, departing significantly from the private property regimes of Europe and America, the laws designated significant fishing grounds for all commoners as a whole to access freely. In Beamer's assessment, from the earliest codification of Hawaiian Kingdom law, " ʻāina was conceived as jointly ‘owned’ through the shared interests of these three classes, thereby reflecting ancient Hawaiian land tenure principles” (2014, loc. 1952 of 5334).

The following year the Hawaiian Kingdom issued the 1840 Kumu Kānāwai, the Constitution of 1840, separating the powers of government into executive, legislative, and judicial branches. The executive branch remained the mōʻī and kuhina nui.[§] The legislative branch was composed of two parts, the first being some of the highest ranking aliʻi and advisers to the mōʻī, similar to earlier structures of governance. The second part of the legislative branch, the Peo i Kohoia, or Representative Body, constituted the biggest change. In theory, this body would be chosen by and represent the will of the people. The Constitution of 1840 also reaffirmed more commons-based land “ownership” to which Kānaka had inalienable rights (Beamer 2014).

Five years after the establishment of the constitution, a Board of Land Commissioners was established to investigate the land ownership claims of those who had previously acquired titles. Years of debate ensued about how to settle joint ownership of lands between mōʻī, aliʻi, and makaʻāinana as codified in the 1839 laws and the constitution. The result was the Māhele of 1848, still a highly contested event among historians in terms of its immediate and lasting results. Until recently the Māhele was largely interpreted as the ultimate and final privatization of land. Complicating this analysis, Beamer and others have argued that Hawaiian Kingdom property laws were a “unique set of hybrid laws created through the blending of Hawaiian tradition and aliʻi appropriation of Euro-American understandings of law” (2014, loc. 3326 of 5334).

The Māhele established distinct land bases for common people, for aliʻi, and for the mōʻī and government. Upon instituting the Māhele, King Kauikeaouli wrote out his intention to give away his and the aliʻi’s rights to land in order to put them fully and forever in the hands of the people: “I hereby give entirely and forever separating the rights of the chiefs and the people of my Kingdom, the majority of my lands so that justice and blessing may come to the Hawaiian Kingdom government” (quoted in Beamer and Tong 2016). In the Māhele land division, the mōʻī’s “Crown Lands” became their property to lease and sell, but also to manage in ways that reserved the rights of Native Hawaiians. King Kauikeaouli set aside the greater portion of the lands he received for “Government Lands,” managed under the direction of the minister of the interior. Lands granted to aliʻi and konohiki were also “koe nae ke kuleana o na kanaka,” or subject to the rights of Native Hawaiians. Aliʻi used

§ Commonly translated as “prime minister,” but more accurately special counselor to the mōʻī with power to veto actions and with authority to execute matters of the government.

these lands in a variety of ways, frequently preserving communal access and connection to ʻāina for makaʻāinana. Some aliʻi created land hui with hybrid collective land ownership; others leased and sold their land to the Kānaka who occupied it (Beamer and Tong 2016).

A central debate during and following the Māhele was how to fairly allocate land claims to commoners. In 1849 a resolution was passed that fee simple titles would be granted to all native tenants "who occupy and improve any portions of any government land" (quoted in Beamer 2014, loc. 2362 of 5334). The resolution also designated Government Lands to be sold at a low price to natives who did not have sufficient lands. The Kuleana Act of 1850 gave makaʻāinana the right to acquire fee simple title to their lands. Relatively few makaʻāinana submitted claims for land, perhaps because outright exclusive ownership was a foreign concept. Andrade (2008) argues that Hawaiians would not have anticipated the potential of eviction from lands they had lived on for centuries. Some historians have hypothesized that commoners may have believed that if they did not act to change anything, the traditional land tenure system would remain intact.

Under the Kuleana Act, eighty thousand commoners secured only 28,658 of 4.2 million acres. However, this figure does not reflect the amount of land that makaʻāinana were ultimately able to secure in the Māhele process. Work by Donovan Preza (2010) has estimated that 167,290 additional acres were acquired by makaʻāinana between 1850 and 1893 through the purchase of low-priced Government Lands. There was clearly attempt on the part of the mōʻī to secure Hawaiian rights to Hawaiian lands. Moreover, the Kuleana Act once again reiterated principles of common land rights: "When the landlords have taken allodial titles to their lands, the people on each of their lands, shall not be deprived of the right to take firewood, house timber, aho cord, thatch, or ti leaf, from the land on which they life, for their own private use, should they need them . . . The people shall also have a right to drinking water, and running water, and the right of way" (quoted in Beamer 2014, loc. 2388).

While not to the same extent as pre-contact production that functioned by universal access to essential fishing grounds, mountain resources, and water, the aim of the laws was in part to maintain and formalize common rights. Beamer and Tong write that the Māhele was "both the formal establishment of private property and the establishment of a hybrid trust and socialistic kind of land tenure that offered special habitation, access, and resource rights to Native Hawaiian subjects" (2016, 131). Significantly, the Māhele—from which all land titles today trace their origin—reserved the vested rights of all Hawaiians to Hawaiian lands for perpetuity.

While the Māhele was not intended or codified as an indiscriminate adoption of exclusive capitalist property, it also did ultimately make large-scale land privatization possible. Euro-American capitalists, missionaries, and settlers in the islands were seasoned in exploiting private profit opportunities. When granted the right to purchase and own lands in 1850, speculative investment in the new land market ballooned. By 1862 it was estimated that three-quarters of O'ahu's land was under control of foreigners (Hasager and Kelly 2001, 195). Correspondingly, Hawaiian commoners increasingly lost access to lands and resources. They were not idle to these changes and expressed discontent in a multitude of ways. Thousands signed letters and petitions to the government and newspapers. They demanded redress against oppressive konohiki or the loss of access rights, an end of land sales to foreigners, and the dissolution of foreigners' influence and presence in government (Osorio 2002; Silva 2004). While some of these calls were heard and addressed in various ways—ali'i forming collective land hui, or government opening their lands for inexpensive sale to Hawaiian commoners, for instance—the privatization of land ultimately hastened the dispossession of Kānaka.

There remains a robust conversation among scholars of Hawai'i as to how to interpret the Māhele and its impacts. Recent scholarship argues not only for recognizing the agency of Kānaka Maoli and their use of tools of the colonizers for their own aims, but also for rethinking the political implications for today in terms of native rights to land (Beamer and Tong 2016). The Māhele process was not a mere imperial imposition naively embraced by Hawaiian rulers. However, new legal and land relations did increasingly trend toward capitalist privatization and markets that dispossessed Kānaka and transferred lands to Euro-American foreigners for commodity production. This transfer of land progressively dug imperialist-capitalist ambitions into the soils of the islands and ramped up the wealth and power concentrating drives of capitalism. While Hawaiians of every class showed remarkable resilience and resistance in the face of immense change and pressure brought by Euro-Americans, the enormous force of capitalism backed by brute military violence is one that reached ever deeper into Hawaiian lands and society.

Rise of the Plantation, Occupation of the Nation

For Euro-American foreigners, the push for privatization of land was never separate from visions of a capitalist economy based on large-scale export agriculture. In her detailed history of the sugar industry in Hawai'i, Carol MacLennan argues that following the Māhele, haole (white) rhetoric of plantation agricultural wealth and the overall health of the Hawaiian nation became inextricably linked (MacLennan 2014, 62). This logic was clearly articulated by William Lee, a primary adviser of the Māhele process, at the opening meeting of the Royal Hawaiian Agricultural Society: "The importance of agriculture and the necessity for its encouragement as a means of national prosperity must be obvious to all. The culture of the soil lies at the bottom of all culture, mental, moral and physical . . . It is an axiom of history too plain to admit the question, that until the savage abandons his roaming, hunting, and fishing, and laying aside his vagrant habits, confines himself to some fixed abode and improves the soil, he can never become a civilized being" (1850, 30).

For Lee and other Euro-Americans, the communal, ʻohana-oriented production system of Hawaiians was backward; only profit-producing agriculture would encourage people to rise above the "immediate wants of their subsistence" (Lee 1850, 32). Again, the racist capitalist ideas of haole were not uncritically taken in by Hawaiian rulers. King Kauikeaouli was negotiating the imperial-commercial pressures of foreign debt, attempting to develop a prosperous modern nation in the face of widespread death, and maintain independence, all at once. Commercial, export-oriented agriculture appeared a viable strategy for navigating these increasing capitalist-imperialist pressures. The king himself initiated the first plantations, making loans available to planters and founding a Royal Hawaiian Agricultural Society to share experience and research (MacLennan 2014). Many different crops were experimented with for their commodity potential, but it was sugar that rapidly became the single dominant industry and transformed the physical and social landscape of the islands.

Commercial sugar plantations were already scattered around the islands by the 1840s, including three ventures owned by the king. These stayed small in scale and played an insignificant role in the first decades of global trade because planters lacked access to labor and land ownership. Besides a small number of Chinese producers who provided milled sugar to Chinese merchants, other early ventures were largely failures (MacLennan 2014). Through this "trial-and-error period," American planters repeatedly drew on the financial support of the Hawaiian government. Carol MacLennan describes this as the beginning of "a lengthy practice that mingled the interests of the sugar plantations with government policy and induced regular cooperation among would-be competitors" (2014, 115). Following the Māhele, missionaries and other foreigners with close ties to the mōʻī had the easiest time buying land, and they increasingly developed personal economic interests in commercial agriculture. However, it was their children who leveraged this historical colonial advantage to reap huge profits.

The American Civil War in the 1860s created a sugar market boom, bringing an influx of capital investment and leading to the development of the first larger steam-powered mills. Many of these plantations were owned and financed by ex-missionaries or their children, often located on lands purchased in the 1850s immediately following the Māhele. At the end of the Civil War sugar boom, planters who were unable to gain financial support from outside capitalists or the government went out of business. However, the short market surge whetted planters' appetites for the sweet profits of sugar and more deeply tied the Hawaiian Kingdom's treasury to its successes. Seeking a reciprocity trade agreement with the United States, planters increased pressure on the government to offer Pu'uloa, Pearl Harbor, to the United States in exchange for access to its sugar market. The vast majority of Hawaiians fiercely resisted the cession of any territory, viewing it as the beginning of loss of all land and sovereignty (Osorio 2002).

Tensions around Pu'uloa, reciprocity, independence, and foreigners' influence were at a climactic point during the most contested 1874 election in the Kingdom's history between Queen Emma and Kalākaua. Hawaiians fighting to maintain sovereignty mobilized in large numbers, alarming sugar capitalists. Reflecting a widely held sentiment, Queen Emma declared, "The natives are all awake now to the American intention of taking possession of these Islands for themselves, and they oppose them to their faces. . . . It has taken the Hawaiian Nation nearly 20 years to learn their Dissenting Missionaries' true character" (in Osorio 2002, 152). Queen Emma vocally opposed reciprocity as a clever "lie" (Osorio 2002, 167). Kalākaua, though considered by the haole elite to be somewhat unpredictable and not entirely cooperative, was seen as at least preferable to Queen Emma. Though both candidates attracted Hawaiian crowds in the thousands, it was through what historian Noel Kent describes as the haole elite's "bribery, threats, and cajoling" (Kent 1993, 45) that the electoral votes were secured for Kalākaua. Riots broke out following his unpopular election, enjoining the participation of the Royal Hawaiian Police themselves. At the request of Charles Bishop (banker, plantation owner, and minister of foreign affairs), American and British troops stationed in Honolulu Harbor occupied the streets and arrested Hawaiian demonstrators.

Against the opposition of the majority of Hawaiians, planters were finally able to secure a seven-year reciprocity treaty with the United States beginning in 1876. While King Kalākaua was successful in keeping cessation of Pearl Harbor out of the treaty, its temporary tenure kept conversations about closer political ties to the US alive. Investment in sugar plantations ballooned immediately following the treaty signing, with forty-two new plantations appearing in the first four years alone (MacLennan 2014). One newspaper described efforts to increase sugar production following the treaty as a "veritable mania" (Beechert 1985, 80). Within days of the treaty's signing, San Franciscan capitalist Claus Spreckels boarded a ship to Hawai'i and secured nearly half of the crop for that year to send to his California refinery. He bought land and water rights, developed the largest and most technologically advanced plantation in the islands, and started an irrigation company and steamship line. While Spreckels himself did not stay in the islands, he left a lasting legacy in his business model of vertical integration that was soon taken to a new level by the Big Five sugar corporations.

Within fourteen years of the reciprocity treaty's signing, ten times as much sugar was being harvested and exported to the US (Kent 1993). Reciprocity encouraged complete dependence on export sugar and the marginalization of alternatives. Noel Kent writes, "In such an encompassing monoculture economy, 'economic diversification' meant at most the production of a few agricultural export crops, like coffee and rice, for the same markets to which sugar was sent" (Kent 1993, 46). Historian and professor of Hawaiian studies Jon Osorio argues that the treaty had the most significant effect on Hawaiian society since the Māhele, benefiting a small class of white capitalists and promoting the "plantation economy over the still viable subsistence of the Kānaka" (2002, 166). The treaty locked Hawai'i into a narrow economic path and dependence on the US, all while delivering tremendous wealth and power to those who would bring the government further under their control. Hawaiian politician Joseph Nāwahī was astute in his observation at the time: "the first step of annexation later on" (Osorio 2002, 168).

Hawaiian rulers through the nineteenth century had sought to develop a legal system and foreign commerce that they believed would ensure independence. They were largely successful in these efforts, masterfully negotiating between and around imperialist powers. Mō'ī and representatives of the Hawaiian Kingdom traveled the world extensively, securing treaties and diplomatic relationships; they managed to use international rules of diplomacy to restore sovereignty when a British consul seized the Kingdom as a protectorate in 1843; they adopted written language and developed one of the most literate nations in the world (Beamer 2014). Most generally, Hawaiian leaders sought to master the protocols and systems of Euro-Americans in order to maintain Kānaka control of Kānaka lands, culture, and government. They were able to preserve an unusual amount of power and independence during a century marked by Euro-American expansionism and colonialism. At the same time, as they adopted modified forms of Western governance and property rights in order to meet the challenges of an imperial world, a foreign capitalist elite seized on the profit opportunities being opened in the islands. This primarily

American elite increasingly found their interests incompatible with those of the Hawaiian monarchy and national sovereignty.

The reign of King Kalākaua brought these tensions to a pinnacle. Though King Kalākaua was seen by the planter class as more amendable to their interests than Queen Emma, he increasingly clashed with their wishes and stirred nationalist sentiment. King Kalākaua toured the world, visiting lands outside of Europe and the United States, and showing a keen interest in cultures and religions of the East. He championed Hawaiian culture, revived study in Hawaiian science (against missionary teachings), and even worked toward a Pacific federation of nations that would protect other Pacific peoples from being colonized. The king's cultural revival, correspondent with calls of "Hawaiʻi for Hawaiians," put him in direct confrontation with missionaries and stirred sentiment that the haole minority perceived as threatening their power and political stability (Silva 2004; Osorio 2002).

Responding to the rise of Hawaiian nationalism and growing resistance to sugar capitalists, in 1887 a small group of haole (the majority of whom were descended from missionaries) founded the Hawaiian League to organize around a new constitution. King Kalākaua was forced by gunpoint to sign the "Bayonet Constitution," terminating nearly all executive power and royal authority and expanding the political privileges of white foreigners. Businessmen entered the legislature in large numbers while Hawaiians' and Asians' suffrage and political rights were severely limited or exterminated by property qualifications and race discrimination (Osorio 2002). Sugar interests were now firmly placed to be the primary adjudicators of the Kingdom's laws and treasury (Osorio 2002, 249). The same year, the Reciprocity Treaty was renewed with the granting of further military control to the US, including the cession of Pearl Harbor.

Within only a few short years, sugar profits were again disrupted by changes in US policy that eliminated Hawaiʻi's preferential treatment, triggering new and more widespread support among planters for complete annexation. Prior to the Bayonet Constitution, a majority of the business elite in the islands opposed annexation, fearing it would stir insurrection and social instability (Osorio 2002, 131). However, conditions changed as the elite was emboldened post-Bayonet and the US indicated that it would require increasing political and military power over the islands in exchange for economic relations favorable to sugar. They were met head on by King Kalākaua's sister and successor Queen Liliʻuokalani, who upon becoming mōʻi immediately began fighting to restore the 1864 Constitution and political power of Kānaka. This brought her into bitter battle with those who had forced the Bayonet Constitution and were secretly meeting with US secretaries of war and state to secure support for an overthrow of the Hawaiian Kingdom.

On January 17, 1893, a small group of haole calling themselves the Committee of Safety orchestrated the overthrow of the Hawaiian Kingdom. Led by attorney Lorrin A. Thurston, the events were entirely made possible by the support of US foreign minister John L. Stevens and the US military. "Eighteen men representing nobody" was a popular characterization of the overthrow (Kent 1993), but those men were backed by the presence of US warships and soldiers, as they often had been through the decades of imperial-capitalist development. The small group of haole elite read a proclamation claiming the monarchy had been abrogated and that they were the provisional government, recognized immediately by Stevens. Queen Lili'uokalani refused to recognize the provisional government and submitted a formal protest to the United States. Shortly after, United States president Grover Cleveland resumed office and ordered Congressman James Blount to investigate the happenings. Following a scathing report from Blount, Cleveland withdrew consideration of annexing Hawai'i and called for the restoration of the Kingdom. The restoration did not take place. Instead, the provisional government renamed itself the Republic of Hawai'i and continued to push for annexation.

Hawaiians mobilized en masse to block annexation and preserve their nation. Large-scale petition drives reached even the most remote dwellers, gaining the signature and formal protest of nearly every Kānaka Maoli (Silva 2004). Two Hawaiian national groups—the Hui Kālai'āina and the Hui Aloha 'Āina—organized meetings with other countries to inform them of the takeover and the discontent of the people. In 1895 there was a small armed uprising attempting to restore Hawaiian rule. As soon as she was freed from imprisonment in the palace, Queen Lili'uokalani traveled to the United States to meet with officials and record her objection to annexation. With a sophisticated understanding of international law, she appealed to other countries not to recognize the illegitimate Republic (Beamer 2014).

With Hawaiians organizing in strong resistance, it was intense militarization and repression that kept order and quelled social uprising in the years following the overthrow. The fraudulent government of the Republic enrolled any willing non-native person to take up arms in their defense. American citizens donated money to buy US weapons, and the government barricaded itself behind a fortress of armory. People suspected of being loyal to the queen were arrested without trial, including newspaper editors charged with publishing "seditious articles" (Beamer 2014, loc. 3167 of 5334). Government employees who refused to swear an oath of allegiance to the Republic were fired.

Despite their best efforts, the fraudulent government was unable to secure an annexation treaty with the United States. Hawaiian resistance was too strong, as was the racism of Americans who wanted nothing to do with a territory of "leprous Kānakas" and "mongrel senators" (Kent 1993, 65). With an equally racist response, sugar lobbyists traveled to the US to warn Congress that if it did not act decisively, the islands would be taken by another "foreign people" in the "white race against the yellow" ("Firmness, Once More" 1897, 4). The *Pacific Commercial Advertiser* appealed to Americans: "There are hardly 2000 of us 'able bodied' men who are trying to hold the fort of white civilization here against 80,000 or more, who oppose us. We need to make our frontage solid as granite" ("Asked to Suggest a Policy" 1898, 4). The pleas of white capitalists in Hawai'i remained in contest for years with American sugar beet capitalists and American racists opposed to a nonwhite nation.

Most finally, it was the ambition of Pacific expansionism that provoked the US to solidify their rule over the islands. When the Spanish-American War broke out in 1898, the strategic importance of Hawai'i was clear. As the *San Francisco Evening Bulletin* editorialized: "[Hawaii is] the center point of the North Pacific. It is in or near to the direct track of commerce from all Atlantic ports . . . It is the key to the whole system. In the possession of the United States, it will give us command of the Pacific" (in Kent 1993, 66). Unable to secure a two-thirds majority in the US Congress for an annexation treaty—in part because of the protests of Hawaiians and the queen—the US resorted to a domestic joint resolution at the height of war. Known as the Newlands Resolution, its legality remains contested today (Sai 2008). Under both international and domestic US law, territory can only be acquired by treaty. The domestic joint resolution is not a treaty and was never voted on in Hawai'i, which in 1898 was an internationally recognized sovereign nation state. Four days after the Newlands Resolution was passed, two thousand United States troops invaded Honolulu, an occupation that remains to this day (Vogeler 2014). The same year, the US took over Puerto Rico, Guam, and the Philippines.

The overthrow and subsequent annexation was a "critical assault on indigenous Hawaiian governance" that broke international treaties and disregarded the laws of nations (Beamer 2014, loc. 239 of 5334). In the analysis of scholars like Beamer, it was the most significant imperial attack on Hawaiian society, followed by loss of Kānaka power, language, and land. Rather than a culmination of colonizing events, it was a conspiracy of sugar oligarchs and American imperialists to break indigenous power. Though Hawaiian rulers' strategic accommodation of Western law and property did open the door to these forces, their maneuvering is also likely what maintained Hawaiian political sovereignty through most of the nineteenth century. Following the overthrow and US occupation, what remained of Hawaiian power was even more systematically assaulted.

The Capitalist State and Sugar's Expansion

Capitalism as a system is relentless in its dynamics of expansion, enclosure, and commodification. However, there is nothing predetermined about its injection into new frontiers or how it shapes them. For capitalist plantation agriculture to become established and thrive in Hawai'i, it required imperial nations forcing their commercial demands on the islands. It also took increasing accommodation by a local state—initially selectively adapted by indigenous rulers in attempt to maintain Hawaiian and national power, and then more violently enforced by a white oligarchy. From the beginnings of land privatization, the Hawaiian state increasingly secured a capitalist order. As wealth and power consolidated under that order—as it does under capitalism—indigenous governance became a threat and an oligarchy backed by American militarism came to replace it.

Without favorable land, water, forest, labor, infrastructure, tax, and trade policies, sugar could not have been profitable on the global market and a sugar elite could not have amassed such wealth and power. Delivering large tracts of land into cane production required government policy to access vast acreage as well as water diversion rights to facilitate higher productivity and expansion onto dryer lands. Early acquisition of fee simple property by missionary families and other foreigners enabled some of the first commercial plantations and contributed to a lasting legacy of consolidated land ownership. In the fifteen years following the Māhele, rapid land sales were fueled by Euro-American speculation about the promises of agriculture and future land values. By the mid-1860s, much of the best Government Lands had been purchased and the government shifted its policy to leasing rather than selling its lands (MacLennan 2014). MacLennan estimates that within only a few decades following the Māhele, most usable agricultural lands were in the hands of plantations and ranches, either held privately or through long-term, low-rent government leases.

The 1887 Bayonet Constitution coup that robbed the mō'ī of its power and disenfranchised Hawaiians led directly to planters' increasing success in gaining large long-term leases. In the subsequent overthrow of the Hawaiian Kingdom, the illegitimate Republic of Hawai'i declared Crown Lands property of the Republic—a massive and blatant theft intended to secure lands for plantation oligarchs. The 1895 Public Land Act was passed, repealing a majority of the Hawaiian Kingdom's land laws and enabling the lease and sale of Crown Lands (Van Dyke 2008). Confiscated lands of the government and crown were renamed "public lands" in an act that Beamer argues deliberately hid the constitutional rights of Native Hawaiians behind nomenclature that continues to this day to suggest they are owned by the general "public" (2013, 14). Three unelected commissioners made all decisions about these lands stolen from the Hawaiian Kingdom and Kānaka. Plantation and ranch acquisitions happened quickly, nearly doubling in acreage to 1.4 million leased government acres from 1890 to 1898 (MacLennan 2014, 262). Whereas under Hawaiian Kingdom law Kānaka Maoli had rights to occupy Government and Crown Lands, the Land Act criminalized "unlawful occupation" and any Hawaiian national refusing to take an oath of allegiance or pay taxes to the illegitimate government was excluded from being able to obtain land (Beamer 2014).

Because the Māhele and the Kuleana Act recognized the inherited interests of natives in the lands of their ancestors, oligarchs and settlers were highly concerned with uncertain, not entirely exclusive land titles. The 1895 Land Act allowed the unelected president of the Republic to "clear up title," stripping rightful ancestral claims to land (Beamer 2014). Following illegitimate US annexation, the Territory of Hawaii similarly adopted a new title system under which a land court could award an original certificate of title, thereby removing Hawaiian claims to the land. With the theft of Hawaiian Kingdom lands and the stripping of native ancestral claims, Beamer argues that it was the aftermath of the overthrow that ultimately most dispossessed Hawaiians and consolidated the interests of the sugar oligarchs (2014).

After annexation, a debate ensued concerning who would manage stolen "public lands." The laws of the Republic were kept mostly intact, though antimonopoly concerns in Congress did reduce the lengths of leases and sizes of land parcels that could be sold. However, industry lobbied against what they deemed "unrealistic" antitrust laws, and continued to renew their large leases. The territorial government also sanctioned systemic evasion of federal law. Thus, despite rhetoric and small actions in Congress to limit sugar's land monopoly, land acquisitions actually grew during the two decades following American annexation and occupation (MacLennan 2014). Acquired at rates such as two cents an acre per year and sizes like Hawaiian Agricultural Co.'s 190,405 acres or Waiakea Mill Co.'s 95,000 acres, it is estimated that over half of all land in sugar production was leased "public" lands (Kent 1993). The significance of this exceptional government subsidy to the lucrative profits of the industry cannot be overstated.

The importance of water as a public subsidy to the private profits of the sugar elite likewise cannot be overstated. Indigenous Hawaiian agricultural systems used sophisticated ʻauwai (ditch) technology to divert streams and irrigate fields. Elaborate methods of timed water diversion ensured equitable access by users. Water was understood to be the source of all life and was revered as a physical manifestation of Kane, one of four major gods. Misuse would result in forms of sanction by the community through the konohiki. After the Māhele, the law became the primary arbitrator of water rights. Early Hawaiian Kingdom law regarded water as a common-use resource and validated the ancient water rights of Hawaiian users. This principle gradually changed as government licenses were granted to sugar companies to divert large quantities of water from distant watersheds. In the 1870s, Maui's Haiku Sugar Company acquired licenses to take water from government lands and build extensive infrastructure across these lands. The case set precedent and was followed by laws allowing eminent domain over land and water for plantation development; it also set the model of thirty-year water leases (Sproat 2011). De facto control of surface and groundwater by the plantations became the norm. After the overthrow and occupation, court decisions through the twentieth century failed to protect Hawaiian customary water rights. It was not until the case of *McBryde Sugar Co. v. Robinson* in 1973 that the courts began to reaffirm the intention of early Hawaiian Kingdom water law to protect the commons.

Without the ability to divert and control massive amounts of water—to the exclusion and expense of other users—sugar in the islands would have dried up and been unable to expand to leeward lands. The first comprehensive irrigation statistics in 1914 show 95 percent of cane lands on Kaua'i and O'ahu were irrigated, as were 90 percent of Maui lands. Irrigated acreage was reported to be twice as productive (MacLennan 2014). Expansion of sugar production following annexation was accompanied by ambitious and dangerous irrigation projects, blasting tunnels through mountains and transporting concrete deep into the islands' forested interiors. Rights to these interior mountain ranges were now provided by the US government. While sometimes scrutinized, no projects were denied and major plantations maintained their water licenses until they closed.

The volume of water sucked from streams and the ground was massive. An estimated 1.5 tons of water were required for each pound of raw sugar (Kahane and Mardfin 1987). The Ewa Plantation recorded in 1926 that it alone drew up to 103 million gallons from its pumps per day. The same year, all San Francisco homes and industry used 80 million gallons per day (MacLennan 2014). Sugar's thirst—or more accurately, the unquenchable thirst of capital to increase the monocrop productivity of the land—pumped aquifers until they were brackish and left entire stream ecosystems dry. Landscapes were remade and human communities displaced as private profit accumulation came to replace Hawaiian systems based on the principle "ola i ka wai ola, ola e kua'aina, life through the life-giving waters brings life to the people of the land" (Sproat 2011, 537).

Planters' access to the forested interior of islands was not limited to water extraction. Early plantations relied on government and private forest lands for fuel, firewood, and timber. By 1880 Hawai'i's forests showed significant destruction above plantation districts (MacLennan 2014). Wild cattle and goats introduced by Europeans in the late eighteenth century had already decimated much forest land; planters recognized that forest decline was causing drought that would impact their crops. Following the overthrow and annexation, the sugar-controlled government established a forest reserve system that plantations relied on to protect mountain watersheds in order to sustain their water use. Funded by the public treasury, the forest reserve designated and fenced off public reserve lands, planted extensive trees, conducted pest research, and experimented with new species. Within ten years nearly a quarter of Hawai'i's land area was in forest reserve, much of it land stolen in the overthrow (MacLennan 2014).

In addition to natural resources, sugar capitalists sought extensive government support in securing labor. Most Kānaka preferred the freedom of their own noncapitalist production while it was an option, incorporating wage labor and capitalist markets only selectively (MacLennan 2014). Initially, Hawaiians who chose to labor for the plantations opted for short contracts or formed work groups for day tasks. They would strike frequently and demanded what haole capitalists deemed "extortionate" wages. Planters bemoaned that "indolent" natives were "retard[ing] . . . the execution of their plans" (Jarves 1838, 70). They advocated that "vagrants" should be forced to work; not for their 'ohana as they always had, but in the capitalist wage economy (Osorio 2002, 120). Increasingly Hawaiians were forced to turn to plantation labor due to changes brought by depopulation, the decline of communal production systems and land rights, new cash taxes, and vagrancy laws. Even with increasing coercion and dwindling options, Hawaiian commoners' cooperation and participation in wage plantation labor remained inadequate to the wants of the growing sugar industry (Beechert 1985).

Sugar interests sought early on to bring in more exploitable people from other lands to labor on plantations. Aiming to gain the support of ali'i, planters presented their obstacle of a sufficient, controllable, and cheap workforce as a problem of (re)population of the nation. They advised that the Hawaiian race might soon perish completely and that loss of sovereignty would be imminent. While likely not entirely convinced by these claims, ali'i were highly concerned with reversing population collapse caused by disease and supportive of seeking immigrants from other countries. In 1855, foreign minister Robert Wyllie sent urgent pleas abroad seeking laborers, with the first groups of Chinese and Japanese contract workers arriving in the 1850s and 1860s. The contract labor system, which continued until annexation, gave legal authority to planters to control the lives of workers. Contracts were strictly enforced by the state, with sheriff pursuits and heavy fines for those leaving the plantations. Without fear of reprisal, managers used and advocated violence to control workers (Beechert 1985). As MacLennan summarizes: "Without government enforcement of contracts through its police and court offices and a companion immigration policy geared toward labor recruitment, the early sugar companies would never have survived" (2014, 121).

Government promotion and subsidization of sugar went beyond facilitating resource use and labor. Following the Māhele, the mōʻī and government directly financed sugar's expansion and dominance. Tax revenues were spent on roads, bridges, harbors, and other infrastructure essential to the industry. Customs revenues were sacrificed in order to secure sugar export to the United States. The public treasury was used to start the islands' first private bank, Bishop & Co., and then keep it going when it failed. Just as they had strategically adopted law and private property regimes, the Hawaiian ruling class likely believed that to maintain independence and Hawaiian power, they would need to develop a more commercial economy and export base. Cultivating the land was an obvious pathway, and they originally supported sugar production enthusiastically, with the mōʻī owning the first plantations. As plantation sugar became entrenched, lucrative, and monopolized by American capitalists, it took on its own power to capture the state and direct its revenues and resources.

Consolidation of Capital in the Big Five

As land, water, labor, and government money were increasingly devoted to sugar or its spinoff industries, a plantation monoeconomy controlled by a US military–enforced white oligarchy developed. From the first sugar boom in the 1860s, smaller growers started to be eliminated by those with access to the large capital required for the latest technologies, milling, irrigation, and transportation. Newly developing and expensive technologies like steam power demanded that plantations grow in size to pay off debts. Plantations with access to capital gained more resources and further access to capital, while others were absorbed in capitalism's consolidation drive. Honolulu agents originally devoted to securing supplies and organizing shipping for plantations developed into bankers, providing loans to planters. As credit dependency and indebtedness became systemic, agents increasingly inserted themselves into operation decisions, gradually becoming owners or shareholders and centralizing authority and ownership of the industry. Observer Ray Stannard Baker remarked in 1911, "the men who really control the plantations live in Honolulu and employ salaried managers to operate the land" (29). Rather than "agricultural and local," these corporate owners were "urban and absentee"—the Hawai'i "aristocrat is a financier rather than a farmer," described Baker (1911, 29).

In 1889, the four predecessor companies to what would become Hawai'i's sugar oligopoly controlled about half of the sugar crop. Within two decades after the overthrow and annexation, the Big Five controlled 94 percent of the crop (MacLennan 2014). In the US-occupied territory of Hawai'i, Baker estimated that from six to ten men "practically dictate the policies of the island sugar industry" (1911, 30). The five companies that successfully consolidated into a powerful, vertically integrated corporate system—Alexander & Baldwin, American Factors, C. Brewer, Castle & Cooke, and Theo. H. Davies—were descended from four missionary families. According to Carol MacLennan's detailed examination of the capitalist class in Hawai'i, many factors played a role in the oligopoly's descent from missionary families: kinship alliances, generational permanent residency in the islands, the pooling of wealth and influence into vertically organized institutions, early acquisition of land, the organization of property into corporations and trusts, and the sugar agency credit system.

Consolidation of political and economic power by missionary-descendent capitalists did involve fiery competition with capitalists like San Franciscan Claus Spreckels, who arrived later. In the 1870s Spreckels built the largest mill in the islands, equipped with the first railroad and major diversion irrigation system, causing other planters to complain about his "tyrannizing monopoly" (MacLennan 2014). However, post-overthrow resource control and political-economic competition between capitalists largely harmonized under the domination of the Big Five interlocking corporate complex. In what Noel Kent calls "Big Five Territory" (Kent 1993, 69), plantations, utilities, shipping companies, railroads, banks, and the main agricultural support industries were either held directly by one of the major companies or were part of an interlocking network of corporate boards, stockholdings, landholdings, and other forms of owner cooperation (MacLennan 2014, 83). Their vertical integration with all major island enterprises—and whole ownership of the California and Hawaiian Sugar Refining Company (C&H) that dominated the western US sugar market—was remarkable even by comparison to other highly centralized sugar producing regions of the world (MacLennan 2014).

The Big Five exerted their economic power in order to extend it, especially by monopolizing key industries and cutting out competitors. They engaged in a variety of tactics to push out sugar producers who were not part of their dominion, including forcing indebtedness and refusing credit from the financial institutions that they also controlled (Kent 1993). In one example, when pineapple plantation owner James Dole changed from shipping with Big Five–owned Matson to an overseas company, his source of credit was immediately cut off. Castle & Cooke promptly took over the company, then proceeded to monopolize pineapple trade globally. Kent calls the Big Five transportation monopoly the "lucrative trump card" that also gave them control over "virtually the entire cargo of freight and passengers moving between the continent and Hawai'i" (1993, 81). By the 1930s, over 90 percent of small retail stores purchased their supplies through the sugar factors (Kelly 1994).

Economic and political dominance worked in a feedback loop that the elite carefully maintained. Famously, attorney general of Hawai'i Edward Dole remarked in 1903: "There is a government in this Territory which is centralized to an extent unknown in the United States, and probably almost as much centralized as it was in France under Louis XIV" ("Big Five" n.d.). Territorial governors were either part of the sugar elite or obedient to it. From the very beginning of sugar in the islands, planters had worked together to secure the interests of the industry as a whole. While there was some competition between capitalists and their relationships to the Hawaiian Kingdom, after the overthrow the industry unified to secure their political dominance. Under the Hawaiian Sugar Planters' Association (HSPA), the industry recruited workers from abroad; centralized management of wages, housing, and social welfare policies; designed anti-union and immigration policies locally and nationally; and marketed a favorable public face to the world to counteract criticism of Hawai'i's labor system (MacLennan 2014). They provided remarkable unity to deal with a growing labor movement, using the courts, violence, and introduction of new immigrant groups to suppress organizing (Beechert 1985; Kent 1993). In 1911, Baker wrote of the HSPA that "[it] is more powerful far [*sic*] than the territorial government; it has well been called the Hawaiian House of Lords" (29).

Workers, Kānaka, and other non-elites were not docile in the face of the injustices of Hawai'i's oligarchical society. However, the institutionalized, imperially enforced power of the Big Five largely shaped the lives of Hawai'i's people and the landscape of political possibility. According to Baker's assessment, "Fully three-quarters of the population of Hawai'i have no more to say about the government under which they are living than the old slaves" (1911, 32). Hawai'i sugar production was able to compete on the global stage precisely because the illegally installed, antidemocratic oligarchic state secured the industry's elite minority interests, maintained extreme class and racial inequalities, and delivered the land, water, and laborers that it demanded.

Monoeconomy Dependence and No Alternative

Critical to the maintenance of any radically unequal social order is the sense of no realistic alternative, of being stuck in a particular trajectory that is impossible to change. As sugar became "king" (as it was popularly called for decades), it displaced and sidelined other capitalist and noncapitalist livelihood possibilities. It created perceived and actual dependence as "the economy" was tied to plantation exports and, more specifically, to the oligarchs profiting from plantations. When problems or challenges to the industry arose, there was often a manufactured sense of no alternative to whatever was best for sugar bosses. For instance, when the contract labor system was criticized for being close to slavery, Samuel Castle (missionary, sugar investor, and politician) responded: "he who tries to throw odium on our system . . . strikes a serious blow at *every interest* in the country, not the planting interests alone, but the coasting, the mercantile and every other one" (1869, 3; emphasis in original). Opportunities for Hawai'i's people were both materially and ideologically reduced to fitting within the plantation agriculture economy.

While sugar was king, other agricultural production was folded into the plantation complex. More than 1 million acres of ranch land, spread mostly at elevations above the cane belt, provided a food source for plantations and were either owned by plantations or by families heavily intertwined with them (MacLennan 2014). Pineapple was encouraged as a homesteading crop to attract white American settlers to the islands, and in short time became a corporate industry dominated by the capital of the Big Five. The entire island of Lāna'i was purchased to establish a pineapple plantation, eventually owned by Castle & Cooke. Indigenous Hawaiian ahupua'a-based taro production that fed an estimated 300,000 people on 10,000 acres in the early 1800s was largely displaced. Many wetlands that had been used to cultivate taro were transferred into rice production to feed cane workers. In the first decades of the twentieth century sugar became too lucrative to dedicate land to rice, and rice was simply imported instead (MacLennan 2014). While the plantation economy did produce some of its own food for workers, the movement away from indigenous Hawaiian society's sustainable self-sufficiency and toward dependence on goods imported from vast distances grew rapidly alongside the export economy.

After the overthrow, as an intensely monopolized economy solidified, social movement pressure for land reform grew. In response, Congress passed the 1920 Hawaiian Homes Commission Act, redistributing 200,000 acres of "public" lands (those seized in the overthrow) back to Hawaiians. Racist island oligarchs pushed for only marginal lands to be included in the program, so as to "rehabilitate" the natives: "Those who contend that Hawaiians ought to have first choice of the highly cultivated lands completely misunderstood the purpose of rehabilitations. We don't want to make the Hawaiians rich we want to make them work. Give these same squatters rich cane land and they would sit on the lanai and strum a guitar or tickle a ukulele while some Japanese did the hoehana in the fields. That isn't what we want, that isn't rehabilitation" (Governor Charles J. McCarthy, in Hasager and Kelly 2001, 203).

Lobbying by the sugar industry ensured that the most agriculturally useful lands remained in their control. While some valuable sugarcane acres were included in the Hawaiian home lands, it was all leased back to plantations (Van Dyke 2008). Characteristic of territorial era policy, the Hawaiian Homes Commission Act concurrently handed an immense gift to plantation oligarchs by removing restrictions on the size of government parcels that could be sold or leased. The entire island of Lāna'i was purchased shortly thereafter for plantation pineapple.

Most generally, throughout Hawai'i's progression toward a plantation monoeconomy, there was much talk about diversifying production and developing a class of capitalist smallholding farmers. In attempts to disperse criticism of monopolistic sugar landholding, assuage growing pressures from the islands' landless majority, and attract white settlers, various "homesteading" policies were pursued. All of these failed as they lacked the necessary government support, were relegated to marginal lands, or were full of loopholes that produced speculation bubbles (MacLennan 2014). With the post-overthrow passage of "homesteading" policies, the amount of land actually homesteaded declined as sugar acres and concentration of control over resources increased (Horwitz et al. 1969). Behind the rhetoric, the common prevailing logic among the capitalist elite—correspondent to their material interests—was that the wealth of Hawai'i lay in the consolidated dominion of the plantation, not in a nation of capitalist smallholding farmers, and surely not in the more communal production of indigenous Hawai'i.

The industry's control over land and resources, and its binding of the entire commercial economy to its own success, required strategic cultivation of workers' dependence on plantations. From early on, planters recognized that more than coercive contracts and the whip were necessary to maintaining an adequate and obedient workforce. In the 1860s, managers experimented with and fine-tuned systems of control over housing, food rations, health care, and debt at plantation stores to ensure compliance with rules and renewal of contracts (Beechert 1985). With plantations often located far from trading towns, the plantation store evolved into a mechanism to "keep them [workers] in debt" (Haiku Sugar Co. plantation manager in 1867, quoted in MacLennan 2014, 139). Food rations were kept low so that workers would be forced to supplement their diet from the store. MacLennan writes that, for many, the plantation was a "cashless economy that greatly restricted their freedom in a world that increasingly demanded money" (MacLennan 2014, 139).

When the contract labor system was abolished, planters were deeply concerned by the loss of their most coercive forms of labor control. They were also alarmed by workers' increasing rebellion. In addition to coordinated use of violence and anti-labor legislation, industry leaders collaborated to design new methods of wage control and mechanisms for ensuring a steady workforce. Aiming to keep plantation workers from seeking alternative livelihoods, in 1919 the HSPA organized the Industrial Service Bureau, which was tasked with establishing a group of "contented people working in the best interests of the plantations" (quoted in Beechert 1985, 180). Following major strikes in 1920, the Bureau worked urgently to offer families various housing, recreational, and educational programs. It was advocated that working-class people should not receive an education beyond what they would need for employment on plantations, including passage of a bill that required high school tuition as a means to deter plantation workers' children from attending.

The industry's strategies of control, coercion, and paternalism were always racialized. In what Andrew Lind (1968) called a "race-making experience," plantations separated people by racial categories that determined where they lived, what work they did, what wages they received, and which sports teams and classrooms they were a part of. Racialization on the plantation also influenced the division of labor in island society more widely. Reinforced by language and cultural differences, the institutionalized dividing of workers was one of many tools used to suppress labor organizing, manage hierarchies, and maintain the existing order.

As the next chapter will detail, sugar production in the mid and late twentieth century moved to cheaper locations of exploit, largely in response to militant interracial worker organizing. However, the legacies of the plantation persist. Hawai'i's corporate tourism monoeconomy, racialized class structure, concentrated power and resource control, and socio-physical geography continue in the trajectory of the plantation economy. All of these factors are integral to understanding the occupation of the islands by the agrochemical-seed-biotech industry today. Moreover, the fundamental condition of agrochemical operations is the ongoing illegal robbery of Hawai'i's political independence and dispossession of Kānaka Maoli by the United States. Hawai'i's place within America, as an occupied territory, but isolated and subtropical, is a determining factor in agrochemical-seed-biotech companies' decisions to locate in the islands. As with the days of King Sugar, a dominant sense of no alternative to today's tourism-military-agrochemical plantations is manufactured through active displacement of alternatives and reproduction of the status quo by those who most benefit from it.

Chapter 5

From Sugar to Monsanto

It was an antidemocratic, oligopolistic system—involving a high amount of cooperation among the local elite and with their allies in the imperial and capitalist core—that kept sugar production so profitable in the islands. In the mid-twentieth century, the islands' capitalist elite began to change their methods of profiteering as a response to both internal pressures and transformations in the global economy. As sugar all but disappeared in the 1990s, the agrochemical-seed-biotech industry rooted itself in Hawai'i's soils, acquiring a range of public supports and subsidies and finding highly suitable, still intact plantation infrastructures, ideologies, and social structures.

The New Monoeconomy

The cost of sugar production in Hawai'i began to grow in the mid-twentieth century due in large part to militant workers' movements (Horne 2011). Worker resistance to capitalist exploitation was present from the very introduction of wage labor in the islands. Hawaiians refused plantation work, broke repressive contracts, and organized strikes around things like equal pay for women. With the coercive indentured labor system and importation of more legally exploitable workers, resistance tended to take the form of impromptu disputes with bosses (Beechert 1985). Following the end of the contract labor system in 1898, worker organizing increased dramatically. Twenty strikes were recorded in the first year alone. Long-lasting strikes organized within racial groups in the 1920s were met with violent response, including the forced eviction of thousands of families from their homes and the death of twenty people in the 1924 Hanapēpē Massacre (Beechert 1985). In the face of murder, brutality, criminalization, and some crushing defeats, workers continued to strike and organize (Jung 2010).

Planters in Hawai'i long used a strategy of separating and pitting racial groups against one another to suppress worker organizing. The International

Longshoremen's and Warehousemen's Union (ILWU)—informed by a Marxist analysis of inequality, capitalism, and racial and class solidarity—explicitly challenged this strategy (Horne 2011). By 1945, the ILWU succeeded in organizing workers across racial groups and in every strategic Hawai'i industry, paralleling the structure of Big Five power. Two great successes in the 1940s—the first involving the closure of thirty-three of thirty-four plantations, and the second immobilizing the Big Five sugar refinery in California—left unionized workers in a strong position to negotiate wages, benefits, and working conditions. Hawai'i workers' ability to force sugar capitalists to share in their profits increasingly made it a less enticing exploit. At the same moment, sugarcane faced increasing competition from sugar beets, a glut of cane in the global market, and a threat to a tariff-free US market with the Sugar Acts of the 1930s to early 1970s (MacLennan 2014).

As Hawai'i plantation workers gained power and sugar's commodity value declined, capitalists many times wealthier than the local plantation elite started investing in the islands as a tourism destination. The Big Five attempted to some degree to keep sugar profitable in the islands and lamented that they "missed the boat in land development" (Lowell Dillingham, in Kent 1993, 119). In fact, however, island elites adjusted quickly to new profit opportunities being opened after World War II with changes in global capitalism and global travel. Most typically this involved deals between "land-rich but capital-poor" Hawai'i corporations, and "capital-rich but Hawai'i-land-poor" overseas corporations that could finance development projects and provide management expertise (Kent 1993, 121). Development boomed and land prices skyrocketed as the number of tourists arriving to Hawai'i went from 429,000 to 2.6 million in the decade beginning in 1963 (up to 10 million today). The market in land drove prices to a point that agricultural production was increasingly less lucrative to landowners. While some elites expressed sentimentality about maintaining plantation agriculture in the islands, Hawai'i's corporations were not immune to capital's coercive laws of competition. As Castle & Cooke stated to the press following the closure of Kohala Sugar Company in 1971, sugar was simply no longer profitable (MacLennan 2014, 276).

As they phased out plantation agriculture and laid workers off by the thousands, the Big Five began to untie themselves locally. Like other US corporations, they multinationalized, spreading to new places and new industries. This was the globalizing trend of capitalism, marked by American capitalists leading the way in opening new markets and securing cheaper locations of production. Corporations like Castle & Cooke expanded their agricultural enterprises to countries with fewer labor and environmental regulations. Within Hawai'i they used their historical land wealth to become residential and resort developers. This spelled the end of Hawai'i-based control of most of these large and expanding corporations (Kent 1993).

Hawai'i today is often contrasted to the backwardness of its Big Five past—a purported break from a white oligarchy to a multicultural capitalism of opportunity. The Democratic Revolution of the 1950s, in which haole elites were challenged by primarily Hawai'i-born Asians, is celebrated as the end of a particular epoch. Prior to 1954, 80–90 percent of territorial elections were won by Republicans associated with the Big Five. Since then, around 80–90 percent of elections have been won by Democrats (Stauffer 2001, 63). Following the switch from Republican to Democratic control, in the 1960s and 1970s some of the most progressive legislation in the US was passed in Hawai'i (Witeck 2001, 38). However, as George Cooper and Gavan Daws document extensively in their book *Land and Power in Hawai'i* (1985), the electoral "revolutionaries" themselves became increasingly invested in Hawai'i's golden day of tourism-development capitalism. They gained power by promising land reforms that would benefit the working class but ultimately promoted land development and real estate deals benefiting those who came to compose the local political power structure (Fujikane and Okamura 2008). Systemic change increasingly gave way to what Kent calls the era of consensus: "There was a growing tone of complacency (amounting to smugness), a feeling that the labor movement had accomplished its essential objectives, that the great struggles were now history" (1993, 137). A chilling anticommunist crusade through the 1940s and 1950s crushed the vestiges of radicalism as the ILWU transferred its allegiance to the Democratic Party. The union turned to tourism as a source of membership, and soon became one of its strongest proponents (Kent 1993).

Ultimately, fundamental structures of inequality from the plantation era were left mostly unchallenged with the turn from sugar to tourism and Republican to Democratic dominance. Immediately following the Democratic Revolution, concentration of land ownership actually grew. By the 1960s, seventy-five landholders (including the state and federal governments) controlled more than 95 percent of Hawai'i's land (Kelly 1994). Most of the sugar elite held on to their land or developed leased lands, with some leases reverting back to the State of Hawaii or the powerful Bishop Estate or Campbell Estate (MacLennan 2014). Though the final plantation shutdowns of the 1980s and 1990s did reorient land ownership and management—sometimes breaking up large landholdings to sell smaller parcels to the highest buyers—these trends increased the price of land, inequality, and lack of land access for regular people (Melrose, Perroy, and Cares 2016). Large landowners continued to command control of water rights, often in violation of the law (Sproat 2011).

Moreover, though contemporary Hawai'i is mythologized as a "multicultural settler state" of racial harmony and equality (Fujikane and Okamura 2008), it remains highly stratified by an ethnicized class structure in ways that Jonathan Okamura argues reproduce the plantation hierarchy (2008, 19). Haole, Japanese, and Chinese Americans occupy dominant political and class status relative to Hawaiians, Filipinos, Samoans, Micronesians, other Pacific Islanders and newer immigrants. Though worker and antiracist insurrection in Hawai'i was remarkable in its initial liberatory wins, these events were also in part absorbed by a neoliberal global capitalism that accommodates multicultural liberalism while leaving structures of racial capitalism and extreme inequality deeply entrenched (Isaki 2008).

Today, Hawai'i is entirely dependent on a vertically integrated corporate tourism economy. It provides cheap labor, natural resources, infrastructure, and other government support in exchange for low-wage jobs and an inflated cost of living—a change in form but not in function from plantation days of past. Rather than to a local oligarchy, profit accumulated from the islands and islanders flows into the coffers of a global elite (Aoudé 2001). The new monoeconomy demands even greater public financial support than did the plantations, including "airports, roads, sewage facilities, new beaches, promotional activities, and a rationalized governmental bureaucracy capable of implementing all of these" (Kent 1993, 122). Tourists are largely drawn to the islands by depictions of a paradise populated by welcoming natives—what Haunani-Kay Trask calls the "prostitution of Hawaiian culture" (1999, 136). Unceasing growth imperatives generate ever more buildings but increasingly drive residents out of their homes or the islands altogether. Tents lining beaches have come to symbolize Hawai'i's "state of emergency" homelessness crisis (Kaneya 2015). As recognized by the *Honolulu Advertiser* in 1993, it is an economy in which "inflation outstrips incomes" (Kelly 1994, 34). The wide disparity between income and the cost of living has in recent years reached unprecedented proportions (Pape 2015a, 2015b, 2015c). Like sugar, alternatives to tourism development are made infeasible by the intertwined compulsions of the capitalist market (especially in regard to land values) and the primary devotion of the public treasury, public lands and waters, and public infrastructure to already dominant economies.

Second to tourism, the United States military forms the other hegemonic economic bloc in the islands. Hawai'i is one of the most heavily militarized places in the world. The military is even more of a flow-through economy than tourism. Its self-contained bases are managed by companies outside of Hawai'i, while military personnel are highly isolated from the local economy (Ramones 2014). The military controls nearly 6 percent of Hawai'i's total land area and more than 22 percent of land on the most populated island of O‘ahu. It pays little to nothing for the resources that it monopolizes. It is also one of the most environmentally destructive enterprises in the islands, responsible for widespread, highly toxic chemical contamination (Ramones 2014). The human and environmental devastation that it reaps globally, largely from its nerve centers in Hawai‘i, is incomparable to any other institution or industry. The agenda of US empire—and the inseparability of its military and capitalist interests—remains the fundamental colonial force shaping life and land in Hawai‘i.

The corporate remnants of Hawai‘i's sugar oligopoly are now nearly all controlled by overseas shareholders and CEOs. Instead of plantation sugar, the State of Hawaii functions largely to secure the interests of transnational capital in a tourism-military monoeconomy. However, dominant structural characteristics inherited from a plantation economy persist. With new ruling forms of accumulation, but in the foundations of its material and ideological past, Hawai‘i's landscape continues to be marked by its consolidated political, economic, and resource control in service to international capital and US empire. It is these lasting legacies that paved the path for the occupation of the islands by the agrochemical-seed-biotech oligopoly.

The Search for the Next Plantation

Sugar production in the islands continued to decline until it all but disappeared in the 1980s and 1990s. During these decades there was much conversation about what was to become of abandoned workers and how to preserve agricultural lands from being lost to tourism and development (DBEDT 2012b; Kahane and Mardfin 1987). Most private landowners did not hesitate in the turn to "inevitable" subdivisions and resorts. As declared by a First Hawaiian Bank economist: "The plantation land on Oahu will be developed inevitably over the next decade. Also inevitable is that most of the farm workers who will be unemployed are middle-aged and older and will have a hard time finding other jobs" (Sahagun 1994). Landowners strongly rejected any government interference in their "inevitable" business decisions. In the first wave of plantation closings, Castle & Cooke sharply rebuked the state for forming a task force for trying to keep sugar alive in the Kohala region: "The implication is that he [Lieutenant Governor George Ariyoshi, who would head the Kohala Task Force] is going to tell us whether we are going to close the plantation or not. He's not going to tell us. We're going to close it" (in MacLennan 2014, 277). Carol MacLennan suggests that this Kohala Task Force in the 1970s "represented the last stand by the State and the union to have a say in maintaining sugar production in the islands" (MacLennan 2014,

277). After investing $6 million, attempting to coax Castle & Cooke to change their minds, searching for new corporate owners, and even entertaining the idea of turning the Kohala Sugar Company into a state-owned enterprise, the task force concluded by giving loans to a few experimental firms that did not survive (MacLennan 2014).

In the next decades, the state continued in various ways to keep the industry alive with subsidies. They talked a lot about better planning, offered to do more studies, and funded job retraining programs for cane workers in the lower-paying service industry (Kahane and Mardfin 1987; Sahagun 1994; De Lama 1994; Kent 1993). There remained a dominant idea that any substantive agricultural predecessor and alternative to urbanization would necessarily be in the plantation mold (Kahane and Mardfin 1987; Plasch 1981). State reports placed hope in discovering substitutions like sugar for ethanol and niche export crops. They argued strongly for increasing public subsidies to sugar and slashing public health regulations to lower "the cost of doing business" in Hawai'i (Kahane and Mardfin 1987, 43). Moreover, it was reasoned that crisis would require sacrifice: "high risk ventures may have to be attempted [and] the community may also become willing to accept activities which it would otherwise oppose because of their negative environmental and social impact" (Plasch 1981, 246).

While government and industry remained ideologically and materially committed to plantation agribusiness, alternative visions were asserted by social movements. It was recognized that the end of sugar in the islands was also an opportunity to go in different directions. Activists pressed for transitions to smaller-scale, diversified agriculture for local consumption as well as for worker-owned cooperatives (Rohter 1992; Bacon 1995). More equitably distributed local ownership and production would reduce import dependencies and build a more resilient and "green" economy (Rohter 1992, 1994). As is typical of response to movements pushing for social change and disruption of the unjust status quo, such calls were branded "unrealistic" by the powerful. Authoritative institutions like the HSPA dismissed what they called a "self-sufficiency myth." They charged that local food would not attract investment, could not compete with cheap imports, and would require only a very small fraction of agricultural lands to saturate the market (de Lama 1994; Sahagun 1994). They neglected the fact that theirs had been an industry profitable and attractive to capital investment due to trade policy, price supports, and multiple forms of government subsidy.

The state did make minimal gestures toward supporting non-plantation food production for both local and export markets. However, without more fundamental systems change and significant redirection of resources, anything of meaningful breadth could not be realized. University of Hawai'i crop scientist Hector Valenzuela describes one example: "When the plantations closed, about 200 farmers were given two acres of land [each] to cultivate, but they weren't given full support. We didn't show them how to farm. So after a few years they gave up" (Mitra 2014). Likewise, worker-owned cooperative initiatives could not wrestle from the state or the banks the larger amounts of land and startup funding required to realize inspired visions (Bacon 1995).

Most generally, the state did see value in maintaining some amount of "diversified agriculture" in the islands. In 1994, more optimistic legislation stated that "the downsizing of the sugar and pineapple industries is presenting an unprecedented opportunity for the conversion of agriculture into a dynamic growth industry" (Act 264, Session Laws of Hawai'i 1994, and codified in chapter 163D, Hawai'i Revised Statutes). As conversations about what would come to fill this space continued, policymakers in Hawai'i and Washington, DC, were pulled into promising, often utopian projections about the future of agricultural biotechnology. State reports predicted that biotechnology would deliver high-value products and high-paying jobs. Local newspapers reported projections of a soon to be $7 billion industry in "crispier and less oily French fries, sweeter fruit with fewer calories, pest- and virus-resistant plants, bigger crop yields and better antibiotics" (Altonn 1998). Governor Ben Cayetano declared that Hawai'i was destined to become the "Silicon Valley of the plant and ocean world" while futurist consultants warned that the state must act quickly if it wanted to attract global biotech commerce (Altonn 1998). More than just agriculture, this was the "lead of high tech," the cutting-edge of the 1990s "New Economy" (Loudat and Kasturi 2009, 7; Aoudé 2001).

Calls for creation of a high-technology sector so as not to miss out on the New Economy corresponded with heightened attention to "economic diversification." Though linked to ideas of reducing economic dependencies, diversification chiefly referred to government incentivization and support of new private capitalist initiatives. It was typically taken for granted that such initiatives would be led and dictated by transnational capital with the islands acting as host. High tech was largely lauded by industry, government, academia, and the media as a sector that would diversify, stabilize, and grow the economy. A piece published in the popular book *The Price of Paradise* declared that standards of living would rise, the "brain drain" would be stemmed, and children would be freed from the destinies of their parents "changing bed sheets" in the service industry (Darby and Jussawalla 1993, 49). A state committee report for a 1999 technology omnibus bill concluded that "high technology promises to be the major industry of the future, fast growing and enhancing our everyday lives in more ways than can be imagined. Your Committee's commitment to fostering high technology growth and development in this State will ensure Hawai'i's prominent role as a mecca for high technology companies and a world-renowned center for innovation and invention" ("Technology Omnibus" 1999). The state created, staffed, and funded numerous programs and agencies to "pursue the potential high tech utopia" (Herbig and Kramer 1994, 58). These included the High Technology Development Corporation, the Hawaii Information Network Corporation, the Office of Space Industry, the Hawaii Innovation Development Program, the Hawaii Strategic Development Corporation, the Research and Development Industry Promotion Program, and others (Darby and Jussawalla 1993).

Despite its investments, commentators chided the state for spending "tens of millions" but "failing to attract even one sizable company" (Herbig and Kramer 1994, 58; Darby and Jussawalla 1993, 45). This led to various pro-capitalist, anti-worker prescriptions. Several papers published by Paul Herbig and Hugh Kramer, both of university marketing departments, aimed to identify and advocate for "factors critical in formation of innovative hot spots." According to Herbig and Kramer, these included a "pro-business attitude," unions being the "exception not the rule," public research institutions that served private business, government subsidy to high-tech business, and a culture that "admired" entrepreneurship (Herbig and Kramer 1993, 111–112; 1994). They warned that "Aloha Spirit" values of friendliness, sharing, and caring for one another might not be conducive to "appreciation of hard work and individualism [that] encourages and provides incentives to be entrepreneurial" (1994, 61). Other academics similarly argued that the state was not sufficiently "pro-business," that it had fallen behind in the deregulatory trend, and that more had to be done to "woo" large multinationals (McClain 1992; Darby and Jussawalla 1993; see various articles in Roth 1992 and 1993).

The search to diversify and bring the "old" capitalist economy into the New Economy through the 1980s and 1990s was notably neoliberal in nature. According to the state, its "bold" policy steps toward a New Economy included deregulation, tax cuts for business, subsidies for technology companies, partnerships between business and government for high-tech growth, and autonomy for the University of Hawai'i to pursue privatizable high-tech research (Aoudé 2001, xxi; see also the New Economy Omnibus Bill 2000). In his 2000 state of the state address, Governor Ben Cayetano boasted that corporations gave his technology omnibus bill "high marks" for being "one of the most progressive in the nation" (Cayetano 2000). Concurrently, record allocations of state funds were poured into tourism, and a special legislative session was convened to authorize the building of a $136 million convention center that would reportedly diversify the visitor base (Kent 1994). These "pump the economy"

(Petranek 2001, 1) and "fiscal emergency" (Witeck 2001, 47) measures were the other half of policy attempts—many successful—at wage freezes, cuts in workers' compensation, erosion of collective bargaining rights, slashes in social spending, and privatization of government services (Witeck 2001, 47; Kent 1994).

Though clearly located in what is widely identified as (new) neoliberal policy, and though seeking to capture a place in the New Economy, in many ways there was nothing particularly new about the prescriptions being ushered in during this period. The state was far more adept and familiar at facilitating transnational corporate investment and activity than critics charged. The biotechnology industry was a primary benefactor of New Economy measures, receiving public subsidies in the form of investment capital and tax cuts (Kanehe 2014; Petranek 2001; Higa 2012). Cayetano was named the Biotechnology Industry Organization's 2002 Governor of the Year, and the state received exhaustive praise: "Gov. Cayetano and the Legislature have demonstrated strong commitment to establishing Hawai'i as a center of excellence for biotechnological research by enacting sweeping economic development and pro-technology legislation over the past several years" ("Biotechs Call Cayetano Their Governor of the Year" 2002).

Most important, the state began moving public lands and waters that legally and morally belong to Hawaiians into the hands of agrochemical-seed-biotech companies. With these lands, the industry inherited valuable infrastructure like roads and irrigation designed for plantation monocrop production. In the name of diversification, preserving agriculture, and encouraging growth of the New Economy, the state rushed to support the next corporate plantations while abandoning any meaningful attempts for something truly new. Native Hawaiian activist and farmer Walter Ritte reflects on the brief post-sugar glimpse at a potential shift, even if slight, in policy trajectory: "Back then the University of Hawai'i's agricultural extension agents would come by and say that we were going into diversified ag and truck farming and that they were going to provide us with the training and support to make that transition. But that never happened . . . All of a sudden the best lands were being given to these big chemical companies and we were back to industrial ag again" (Mitra 2014).

The Plantation Social Landscape

When the agrochemical-seed-biotech industry expanded their operations in the islands, the former sugar and pineapple lands that they occupied surrounded communities created and subsequently abandoned by the plantations. Many of these communities are isolated from tourism centers, located in places where lost jobs, housing, and health care are not easily replaced. Moreover, the hard-won wages of unionized plantation workers were far higher than in today's resort industry. In places like Moloka'i, communities actively resisted the replacement of agriculture with tourism. Walter Ritte describes it thus: "On Molokai we fought tourism so we could keep our subsistence and our way of life, our safety net. But that meant we supported ag. They [the state] told us diversified ag would come in, but then these guys [the chemical industry] did. And nobody knew what it was at first. It took a long time before we figured out what they were doing" (personal communication).

The communities that the industry moved into are almost entirely working class, with higher concentration of Hawai'i's most marginalized racial and ethnic groups. They face systemic underemployment and wealth and power disparities that are continuations of colonial and plantation hierarchies (Okamura 2008). The west side of Kaua'i, for instance, has the highest percentages of households receiving food stamps and children living in poverty compared to the rest of the island (JFFG Draft 2016, 70). It is also host to the island's overflowing landfill, several highly contaminated toxic waste sites left by sugar, and the world's largest missile testing range (Siegel 2011). As one Native Hawaiian woman from the west side put it, "it's no coincidence that they [the industry] came here."

Racialized plantation hierarchies are largely transposed onto today's fields, with many fieldworkers and managers having transferred directly from sugar or pineapple. "Same bodies, different boss companies," as one resident put it. Nearly all fieldworkers are from Hawai'i's most marginalized ethnic and racial groups, including Filipina/o, Native Hawaiians, and migrants from Southeast Asia, Pacific islands, and Latin America (Shaw 2016a; Hofschneider 2015b). Given Hawai'i's occupational inequalities, the wages received in agrochemical fieldwork are nominally higher than what would be earned at the bottom rung of the service industry (Loudat and Kasturi 2013; Okamura 2008). However, according to the industry, only around half of their 1,150 jobs are full-time (Loudat and Kasturi 2013). Companies offer a small number of higher paying managerial roles to local people (which also serves to secure local allegiances), but most senior management roles are filled by white men who are not from Hawai'i (Shaw 2016a).

While the industry represents job creation as a significant contribution to Hawai'i's economy, it also speaks privately about using extensive migrant labor because "local people don't want these jobs" (labor recruiter, pers. comm.). Residents on Kaua'i allege that they were initially hired by companies but laid off in waves as they were replaced with migrant workers. Third-party contractors typically bring in migrant work crews for 8–12 weeks during the busier winter and summer pollination and harvest months, then move these workers to US continental locations. A severe lack of public information and regulatory oversight exists regarding the conditions of migrants arriving through temporary guest-worker schemes, which by their very design deprive and undermine basic labor and human rights (Southern Poverty Law Center 2013). Hawai'i's agricultural sector is especially notorious for widespread abuse of migrants. In 2010 it was made public that hundreds of trafficked Thai people were working in conditions of "modern-day slavery" in Hawai'i fields. Guilty agribusiness companies (some that are also the highest users of pesticides in the islands) were strongly defended by two previous governors, the Hawai'i Farm Bureau, the Hawai'i Department of Agriculture, the former head of Hawai'i's Board of Land and Natural Resources, the head of the Hawaii Foodbank, two banks, and many other power holders. Agrochemical companies were not directly implicated in the 2010 Thai worker cases, but they benefit immensely from policy and power structures that enable horrifying exploitation of migrant workers.

Though there is no publicly available data about the composition of Hawai'i's laborers, women appear overrepresented in many agrochemical-seed-biotech industry fieldwork tasks (Shaw 2016a, 2016b). For example, soy pollination is heavily feminized, with labor performed almost exclusively by older Filipina women. Soy pollination is performed under the direct sun, bent low to the ground, using tweezers to move pollen. Managers and labor recruiters tend to refer to workers in highly ethnicized and gendered terms, considering some "hard working" and others "lazy" (Shaw 2016a). Some of the very same racial and gender constructions that existed through the plantation era are reproduced in today's agrochemical fields (Beechert 1985).

Beyond jobs, the agrochemical industry establishes material dependencies in the economically marginalized communities in which they operate. They fund social welfare where the government has abandoned people, sprinkling money at public schools, health centers, domestic violence services, food banks, conservation work, housing projects, sports teams, community gardens and recreation. They privately solicit groups that have wide community support with offers of small grants[¶] for their work in areas described as food security, environment, youth support, and empowering women. They also make subleases available to local farmers and donate parcels of land for agricultural parks. Charity serves to secure allegiances, blunt resistance, and distract from the fundamental question of how and from whom such large sums of capital

¶ The author has seen private email offers mostly in the range of $1,000–2,000.

are being extracted in the first place. Plantations of the past were also remarkable in this regard. As Kent observes of the sugar oligarchs, "It was a ruling class which retained enough of the missionary spirit to delight in lavishing money upon its favorite charities," while at the same time violently refusing policy that would guarantee better working and living conditions (1993, 83). While not to the same extent as sugar, agrochemical companies similarly disperse charity to a range of programs and services that are relied upon by Hawaiʻi's working-class communities while fiercely opposing policy changes that would better people's lives.

Though the agrochemical industry's reach across the landscape is not close to replacing sugar's, it is largely credited with preserving rural agricultural lands and livelihoods. As the *New York Times* reported the dominant narrative: "The firms have spared farmland that would otherwise be lost to development" (Voosen 2011). Or in the words of former mayor of Kauaʻi Bernard Carvalho, this is the "kind of agriculture [that] really feeds our families" (Mitra 2014). The prevailing idea that transnational agribusiness which extracts and exports wealth from the islands is the only economic form that can "feed our families" assumes and obscures numerous forms of public subsidization, particular arrangements of resource control and power, and government facilitation of corporate monopolies and dominance. Of course, most of these conditions originate far beyond Hawaiʻi's shores, in the contours of global monopoly-finance capitalism. A range of international and national policies facilitate unprecedented corporate power in the agrifood system and undermine more equitable, regenerative, regionalized food production. However, as the next chapter will detail, much is also determined by local social relations and decisions, including extensive public supports and subsidies.

Chapter 6

The State's Redistribution of Collective Wealth

For capitalist plantation agriculture to become the established norm in Hawai'i and displace the sophisticated production systems that preceded it, it took both imperial nations forcing their commercial demands on the islands and the backing of a local state. Without favorable land, water, forest, labor, infrastructure, tax, and trade policies, sugar could not have been competitive or profitable on the global market. Likewise, corporate tourism could not have become dominant without massive government investments and the giveaway of island resources to overseas capitalists (Stauffer 2001, 94). Similarly, contra the oft-repeated assertion that agrochemical-seed-biotech industry "contributions . . . are at no cost to the State" (Loudat and Kasturi 2013, 4), their operations are conditioned upon a range of public supports and subsidies, including policies that enable externalization of health and environmental costs onto the public. Much of this support is in the foundation of a plantation past, with sugar's legacies and champions continuing to facilitate the new agribusiness barons. Inseparable from backing by local government and power holders, it is most especially Hawai'i's subjection to US regulatory regimes that makes the islands most attractive to the agrochemical oligopoly.

Land

Land acquisition by agrochemical-seed-biotech corporations is made possible by a continuing history of consolidated land control, including the State of Hawaii's management of public lands seized from the Hawaiian Kingdom. When the indigenous Hawaiian Kingdom was in power, Government and Crown Lands were managed in trust with special rights for Hawaiians in perpetuity. Still contested and being reclaimed today, these lands were stolen in the overthrow. Upon statehood in 1959 their administration was transferred to the Hawai'i state government. Like sugar, much of the acreage occupied by the agrochemical industry today is made up of these so-called ceded lands that were never actually ceded. On the island of Kaua'i, around half of the acreage the industry occupies—at least six thousand acres—are these stolen public lands (JFFG 2016, 21).

Public lands on the island of Kaua'i have been leased to companies through the Agribusiness Development Corporation (ADC), a "public-private corporation" formed during sugar's closures to convert agriculture into a "dynamic growth industry." In the name of bypassing "bureaucratic red tape," the legislature granted the ADC broad powers to buy, own, and develop public lands and waters. It was granted exemptions from public land trust regulation, Public Utilities Commission regulations, and civil service laws. Without clear mandate as to how to serve its purpose, in practice the ADC has functioned to redistribute public resources for narrow private gain through a nondemocratic process in which decisions are made by governor appointees (HDOA 2016). The Hawaii state auditor concluded in 2021 that the ADC "has done little—if anything—toward achieving its statutory purpose" (State of Hawaii Office of the Auditor 2021, 2). The ADC acts with what the islands' leading newspaper calls an "unusual lack of accountability," including failing to submit annual reports to the legislature as required by law. The most recent report posted on its website is over a decade old (Yerton 2018). According to the state auditor, neither the ADC nor its board know what the agency's legally required duties even are. Between 2014 and 2018 the legislature gave more than a quarter billion public dollars to the ADC for operations and capital investments, including for projects that are reportedly displacing communities that have lived and grown food on lands for generations. Lawmakers themselves have not been able to gather basic information about how these large sums of the public treasury have been spent (Yerton 2018).

The ADC controls some of the most fertile agricultural lands on Kaua'i, which are equipped with irrigation and road infrastructure from sugar days. A large majority of these public lands are leased to agrochemical companies, while local farmers seeking leases have been denied without any justification. There is no transparency as to how tenants are selected or terms of leases are negotiated. Leases to agrochemical companies include 20-to-35-year agreements, for rates as low as $50 per acre per year for tillable acres and $1 per acre per year for non-tillable acres (State of Hawaii Agribusiness Development Corporation Agreement No. L-08202, 2007). Companies have lobbied for some of the acreage they occupy to be reclassified as "non-tillable" after having had the land in cultivation. The ADC's most expensive lease is to BASF for $620 per acre per year for a ten-acre parcel that is used for offices and facilities. In 2019, on the west side of Kaua'i the State's Public Land Trust Information lists a lease with Syngenta for 2,369 acres ($267,941 per year total), with BASF for 977 acres ($146,831 per year), another with BASF for 10 acres ($6,200 per year), and one with Pioneer for 1,462 acres ($219,550 per year). These have transferred to Hartung Brothers, Beck's, and Corteva, respectively. Seed companies also sublease an unknown amount of ADC controlled land from other agricultural enterprises, so these numbers may not constitute their total public land use. As described in chapter 3, there has been recent reduction in the amount of land occupied by the industry due to both global/local consolidation and local resistance.

Around 85 percent of land that the industry occupies is leased, from either the state or one of Hawai'i's largest landowners (Freese, Lukens, and Anjomshoaa 2015). Again, consolidated landholding born from plantation days makes this acquisition by the industry possible. Some of these large landholdings are still in original private family trusts descended from plantations, such as the Robinson family. Others have been bought and sold in large concentrated holdings to newer investors. For example, Grove Farm owns 38,000 acres on the island of Kaua'i that were bought immediately following the Māhele for speculative purposes. According to Grove Farm's website, "Warren Goodale became the first owner in 1850. He immediately sold the land to James Marshall for $3,000, who in turn, sold it to Judge Herman Wireman for $8,000 in 1856." These lands were part of the plantation complex until sold to billionaire Steve Case, co-founder of AOL, in 2000. In sum, while some of the Big Five sugar corporations and missionary descendant families no longer own land in Hawai'i, plantation patterns of consolidated land control still shape current land use and acquisition by the industry.

Notably, Kamehameha Schools, the largest private landowner in Hawai'i, leases over 1,000 acres to Monsanto (now Bayer). The lease was originally to Holden's Foundation Seeds in 1999, but it passed over to Monsanto when they acquired the company four years later. Kamehameha Schools is among the wealthiest charitable trusts in the world. It was established in 1884 by the will of Princess Bernice Pauahi Bishop for the purpose of educating and supporting Native Hawaiians. Many Hawaiians contest that this mission runs contrary to their business partnership with the agrochemical-seed-biotech industry (Yap 2013).

For actual farmers and ranchers, access to affordable, long-term land leases is the biggest barrier to agricultural production. This is foremost due to capitalist market pressures to get the highest exchange value out of land through development for tourism, speculative investment, or sale to global elites who pay tens of millions of dollars for vacation homes. With land a rare and astonishingly expensive commodity in Hawai'i, the agrochemical-seed-biotech industry's ability to acquire prime agricultural lands not only from private landholders but also from the state is of major significance. On occasion local farmers and ranchers have been evicted from their leases to make way for the transnational agrochemical-seed-biotech corporations, which are deemed by both private landowners and the state to be more "stable" lessees (Kaua'i rancher, personal communication, 2013; Bynum 2013). In response to criticism, and as a potent strategy of greenwashing and securing local allegiance, the industry has more recently turned to subleasing lands that they control back to a small number of local growers. While significant for those livelihoods, the scale of industry land acquisition remains exceptionally high when compared to other agricultural users.

Water

Water is essential to agrochemical-seed-biotech operations, and large tracts of public and private agricultural lands are already equipped with irrigation infrastructure and political control over water. Control over freshwater was fundamental to the expansion and profitability of sugar. The industry would have dried up without the government enforced ability to divert massive amounts of water out of forested island interiors and away from other ecological and human communities. Water rights were integral in determining who profited and who was dispossessed, and what types of agriculture flourished and what types disappeared. This remains the case today.

The Hawai'i Constitution, State Water Code, and common law protect freshwater as a public trust resource. Contemporary legal recognition of water as a public trust is a product of social movement struggles throughout the 1960s and 1970s to reverse a century of plantation monopoly over water (Sproat 2009, 6). During these years, the Hawai'i Supreme Court ultimately reaffirmed that Hawai'i's water resources are held in trust and should be managed for the benefit of all (Sproat 2011). Court decisions were followed by the adoption of the public trust doctrine at the 1978 Constitutional Convention. The Hawai'i Constitution currently includes the directive: "to protect, control and regulate the use of Hawaii's water resources for the benefit of its people" (State of Hawaii Constitution, Article XI, § 7). Environmental protection, traditional and customary Native Hawaiian rights, appurtenant rights (rights that attach to parcels of land), domestic water uses, and reservations for the

Department of Hawaiian Home Lands have priority over private commercial water uses.

Although Hawai'i state law recognizes water as a most essential environmental and common good, implementation and enforcement of the public trust is severely lacking. Moreover, almost every year legislative attempts are made to erode the public trust purpose of the law. Control over water is still very much tied to Hawai'i's most powerful business and landowning interests (Sproat 2011). It has taken lengthy and expensive citizen-initiated litigation to implement the law in select localized places (Sproat 2011). Agrochemical companies benefit from and rely upon these arrangements. Their use of water is premised upon infrastructure and political decisions that continue to keep water flowing toward wealth and power.

In one primary example, on the west side of Kaua'i the State Water Commission "merely rubber-stamped" historical diversions of the former sugar plantations when management transferred to the Agribusiness Development Corporation (Henkin and Moriwake 2013, vii). The irrigation systems being used by the industry on these public lands were constructed by the Kekaha Sugar Company in the early 1900s after it impaired its groundwater wells from overuse (Henkin and Moriwake 2013, 14). Comparable amounts of water continued to be diverted by the agrochemical industry, though it cultivates only a fraction of the land that was in sugar, and in much less water-intensive crops. Residents, including Native Hawaiian rightholders, allege that both streams and other users are being denied legally due water allocation. Many testify to all-time lows in river water and corresponding ecological degradation. A 2013 legal petition by environmental and resident groups accused the Agribusiness Development Corporation and its tenant Kekaha Agriculture Association (Dow, DuPont, Synganta, and BASF) of "committing unlawful waste, including outright dumping of diverted river water" (Henkin and Moriwake 2013, vii). Due to the persistence of community watershed activists, after four years of negotiation with the ADC and others, an agreement was reached in 2017 to restore water to the Waimea River and downstream users. Under the agreement, tens of millions of gallons of water are supposed to be returned to the river each day, bringing back continuous flow from mountains to sea. The deal signifies the first time that water has been "voluntarily" returned through mediation versus litigation. At the time, Earthjustice water rights attorney Isaac Moriwake hoped it would be a "historic breakthrough" (2017). However, there is concern that in 2021 water has still not been restored to the agreed-upon levels.

Similar plantation-era diversions persist throughout Hawai'i, subsidizing entities like the agrochemical-seed-biotech industry. On Moloka'i, for instance, the industry primarily leases land from Molokai Ranch, which has diverted water out of watersheds to feed plantation crops for over a century. Molokai Ranch owns more than one-third of the island, a legacy of consolidated monopoly plantation landholdings. Like entities in other places throughout Hawai'i, the ranch—and by extension its lessees—is facing legal challenges for continuing unlawful water diversions that residents allege are damaging stream life, aquifers, and Hawaiian fishpond systems. These diversions continue even while Moloka'i is currently designated to be in "severe drought." Precious, contested water is understood to be "the key to Molokai's future" in the words of Kapua Sproat, director of Ka Huli Ao Center for Excellence in Native Hawaiian Law (Hiraishi 2018). Or in the words of Lohiao Paoa, a Native Hawaiian water rights advocate, "returning the water to Kawela ahupuaa [watershed] will bring back life that it once had before" (Uechi 2019).

Taxes and Other Direct Subsidies

In addition to invaluable public land and water, agrochemical-seed-biotech companies are directly subsidized through high-technology tax credits, property tax breaks, taxpayer loss of general excise tax revenue, and unpaid liabilities to the public (Kanehe 2014; Hooser 2015b; Redfeather 2012). In 1999, the state began offering various incentives to "high-technology industries" through the passage of Act 178, followed by Act 221 in 2001 and Act 215 in 2004. Act 178 contained eight initiatives, with tax credits being its hallmark. They sunset in 2010. According to an audit in 2012, these high-technology tax credits cost the state an estimated $1 billion in lost tax revenue (Higa 2012). While at least twenty-five other US states offer comparable high-tech investment tax credits, Hawai'i's was far above the level offered in any other state (Higa 2012, 14). For most of the years these subsidies were in place, the names of the companies receiving them were not made public or revealed to the legislature, unlike in other states where disclosure is mandated (Curtis 2014; Higa 2012). Thus, it is impossible to know which of the agrochemical companies received subsidies and to what extent. Virtually none of the tax claims were audited (Higa 2012). The Biotechnology Industry Organization did give the state sweeping praise for its "support of investment incentives for the Hawaiian biotechnology industry," which it advocated should serve as a "model for other states wishing to promote innovative research and development" ("Biotechs Call Cayetano Their Governor of the Year" 2002). According to the subsidy tracking of the policy research center Good Jobs First, "for its size, Hawai'i gives away an enormous amount in subsidies" and has notably "lavish tax credit programs" (n.d.).

Monsanto and Dow, and likely other companies, also participate in Hawai'i's Enterprise Zones Partnership Program. The program "was established by the Legislature in 1986 to help stimulate certain types of businesses such as biotechnology, information technology, and agriculture" (DBEDT 2007, 2). Eligible businesses are exempt from Hawai'i general excise tax and can claim partial personal or corporate income tax credits for up to seven consecutive years. Counties may offer additional tax subsidies in designated enterprise zones (DBEDT 2012a). Again, due to lack of public disclosure, it is unknown how much companies have privately profited from these public subsidies; given the industry's high praises, the subsidies have likely been significant.

The agrochemical-seed-biotech industry benefits from more disguised tax subsidies as well. Unlike other forms of agricultural production in the islands, the seed development conducted by companies does not generate a product that is then sold. Thus, it excludes payment of the State of Hawaii's 4 percent excise tax. These are lost public revenues that have occurred with the transition from agricultural production to seed-biotechnology research and development (Bynum 2013). At the same time, companies take advantage of agricultural property tax and utility subsidies through land use dedication policy designed to incentivize agricultural production (Kaua'i County Council Bill 2546). Adding to these various forms of subsidy, council members on the island of Kaua'i allege that numerous violations of county law by companies and the landowners who lease to them have resulted in what is likely millions of dollars in unpaid liabilities (D'Angelo 2014).

Research and Education Institutions

Public subsidy and support for private agrochemical-seed-biotech corporations is also funneled through research and education institutions, especially the University of Hawai'i's College of Tropical Agriculture and Human Resources (CTAHR). Like other public research institutions today, the University of Hawai'i (UH) is expected to function as a business, be responsive first and foremost to the capitalist market, and vigorously pursue partnership with private companies. Governor Neil Abercrombie's (in office 2010–2014) "A New Day in Hawaii" campaign stressed that "UH and the private sector must work together to form the economic driver that this state has talked about for decades. Closer collaboration will help transition millions in federally funded research and development activity into products and services that can lead to the startup and success of world-class tech companies" (Enay 2011).

As a public land-grant institution, CTHAR's stated mission is "to support tropical agricultural systems that foster viable communities, a diversified economy, and a healthy environment." In the neoliberal era, this broad public mission is assumed as best carried out through and with the private sector. Most faculty within CTHAR are expected to develop and privatize intellectual property, generate revenues, and serve private business (Redfeather 2012). As CTAHR puts it, their mission is served by "help(ing) all of our *clients*," with the agrochemical-seed-biotech oligopoly taking preference (emphasis added). This prioritization of profit generating research directs public resources away from other research goals and broader social aims. An open letter to the UH chancellor and dean of CTAHR, signed by sixty University of Hawai'i faculty, reads:

> Instead of supporting local production and raising the percentage of homegrown food, CTAHR has increasingly devoted its resources to serving the interests of biotechnology. At times, as many as 60 or 70 CTAHR faculty and staff have been working on biotech projects, despite the fact that, after the initial success of the Rainbow papaya, not one of CTAHR's more than a dozen attempts to produce a commercially-viable genetically engineered plant species has proved successful . . . It is surely questionable whether, in terms of a land-grant institution's mission, such corporations should be treated as members of "Hawai'i's food and agricultural system" at all. Yet CTAHR does not merely regard them as an indispensable part of "diversified agriculture", it has given its full support, in articles, workshops, and outreach programs, to the various activities of these corporations. Time and energy used in such ways has inevitably detracted from time and energy that could have been spent in support of Hawai'i's food producers. (Caron 2015)

In addition to directing public research dollars toward already dominant private industry, CTAHR also trains students in the skills they will need to provide labor for the industry. The agrochemical-seed-biotech industry is the largest employer of agricultural scientists with advanced degrees in Hawai'i (Schrager and Suryanata 2018). Scholarships from the industry—including a Monsanto Research Fellows Fund (started with $500,000 in 2011) and Monsanto Scholarship Fund (started with $100,000 in 2010)—influence the development of these education pathways and students' choices to participate.

Critically important education and research that lacks backing from big capital is sidelined (see also chapter 2).

CTAHR also plays a critical role in shaping public discourse and opinion around agricultural biotechnology in ways that conveniently erase the role of the agrochemical oligopoly. Partly funded by Bayer-Monsanto, its extensive outreach programs focus on feeding the world through unrealized imaginations of technological possibilities. A "DNA Rap" on its website goes:

> It's DNA that is the key
> To finding better crops
> That will make farming easier for me!

In its Gene-ius Day Program for elementary school children, a comic book tells the story of child heroine named Caitlin suddenly unable to find food—"all the farmers took a vacation cause it can get hard to feed the whole nation." Hungry Caitlin is saved by scientists who find the "answers . . . in DNA!" by creating "plants that won't take abuse from wind or floods or dry drought weather." In a happy ending, Caitlin "gets these farmers and scientists to all work together" so that there is enough food. This example is not unusual but indicative of the college's Biotechnology Outreach Program. The program disseminates narratives not only about the purported benefits, but indeed the "necessity" of biotechnology in ways that occlude its actual control and use.

CTAHR faculty who have raised questions about the agrochemical industry or biotechnology have faced various forms of harassment, silencing, and restrictions of their academic freedom. One professor of crop science who publicly expressed concerns about genetically engineered crops was accused by colleagues of "actively supporting the poisonous activities of groups basically opposed to CTAHR, science and progress," "intellectually dishonest arguments," and "insulting to our organization and several of our clients" (Koberstein and Murphy 2015). Such derision suggests institutional suppression of scientific inquiry and debate. Further, it indicates the unquestioned assumption of close relationship between industry and the university. While partnership with industry is expected and encouraged, working with environmental or social justice groups is reproached as leading to "bias" (Koberstein and Murphy 2015). Some voices from within CTAHR are privileged and seen as representing the institution (scientists leading the Biotechnology Outreach Program, for example), while others are marginalized and must publicly declare that their views are separate from those of their institution, even while both are in fact participating in a controversial scientific and social debate.

More generally and beyond CTAHR, the biotech industry has inherited sugar's infrastructure of both public and private research institutions. The University of Hawai'i, the Hawaii Agriculture Research Center (HARC, previously the Hawaiian Sugar Planters' Association), and the more recent Pacific Basin Agricultural Research Center are notable for their transgenic crop research, much of it federally or state funded (USDA 2016). The root problem here is the disproportionate devotion of public research dollars and institutions to technology pathways that are serving very narrow private interest and gain. Public resources are going to bolster an oligopolistic industry that is actively doing social and environmental harm, while resources are being withheld from research and education that are essential to an ecologically regenerative, healthy, and socially just food system.

Unaccounted-for Subsidies in Public Policy

State support for the agrochemical-seed-biotech industry extends beyond land and water, direct financing and tax breaks, and research and education. It also includes policy regimes that privatize benefits and socialize costs more generally. This is a problem with capitalism as a whole. The process of accumulating private profit wholly relies upon things that are not paid for by private capitalists (public infrastructure, nature, social welfare programs that keep underpaid workers alive, science and technology, etc.). At the same time that private profit is extracted from collective goods, costs of production are externalized onto the public (toxic emissions, landfills, atmospheric changes, pollinator death, medical expenses, etc.). With the agrochemical industry in Hawai'i, specific local policy regimes force the public to pay in various ways for industry operations that faraway company shareholders reap the financial rewards of. Just some of these costs include those of human health, psychological trauma and stress, decreased stream flow, contaminated livelihood resources such as beehives and fisheries, depleted soils, pesticide-laden dust in homes, and more. Costs to the public are not evenly distributed; they primarily burden working-class communities of Hawai'i's most marginalized racial and ethnic groups.

Relative to other US states, Hawai'i does possess a decent legal framework for protecting the environment, which was largely enshrined as sugar began to decline (Sproat 2011). In 1978, a constitutional convention elevated the public trust doctrine to a constitutional mandate. The state has a legal duty to hold and utilize all "public trust resources," including "land, water, air, minerals, energy sources" and all flora and fauna, for the benefit of all people (Gibson 2014, 258; Hawai'i Constitution, article XI, § 1). Further, the constitution requires local county governments to do the same. However, "the public trust doctrine tells the State *how* it should make decisions, as opposed to *what* those decisions must be" (Gibson 2014; 258). In the case of regulating agrochemical-seed-biotech operations, as with many others, the constitutional mandate is largely being neglected. The state's policies and *modus operandi* related to pesticide use and big agribusiness are mostly left over from and reminiscent of sugar days. While social movement pressure of recent years has forced some increased regulation of big agribusiness, thus far it has been grossly insufficient to the task of forcing the agrochemical industry to internalize even the worst excesses of their true costs of doing business.

The most controversial element of the current policy landscape is the state's preemption—or invalidation—of counties' rights to regulate agrochemical operations. Significant attempts to protect human health and the environment from impacts of agrochemical operations in three of four island counties have all been nullified by a federal court's determination that it is the sole authority of the state to regulate agriculture. In ruling the County of Kaua'i's pesticide disclosure and buffer zones Bill 2491 invalid, a US District Court judge issued the opinion: "the State of Hawaii has established a comprehensive framework for addressing the application of restricted use pesticides and the planting of GMO crops, which presently precludes local regulation by the County" (Hofschneider 2014a). At the time of the ruling, that "comprehensive framework" included none of the basic pesticide protection laws existing in many other states, such as no-spray zones around population areas, regulation of pesticide use on or near schools, poisoning surveillance programs, and notification requirements for pesticide application (Freese, Lukens, and Anjomshoaa 2015).

Over the past decades, dozens of pesticide bills have been introduced at the Hawaii State Legislature. Most are never scheduled for a first hearing and thus die without legislators ever even voting on them. Until 2018, only one—Act 105—was passed in what Senator Mike Gabbard described as an "extremely watered down" version of the original bill (in Cocke 2013b). The course of Act 105 is illustrative of the general trend around attempts to regulate large agribusiness at the state level. Initially the bill (HB 673) sought to require an annual list of all pesticides used in each county by type and volume, a summary of health complaints related to pesticide use, an analysis of trends in pesticide use, and an assessment of the accuracy of the reported data. Following extensive lobbying by the industry, the Hawai'i Farm Bureau, large landowners, the Hawai'i Department of Agriculture (HDOA), and multiple other state departments, the bill was eviscerated of any meaningful content. What passed was Act 105, mandating the HDOA to post on its website already existing sales records of restricted-use pesticides only. With no start date and no requirements for the timing of postings, the HDOA did nothing. Two years later, pressed by journalists, the HDOA's pesticide branch said that the law had not been implemented because they did not want to be sued by the industry over release of confidential business information: "We wanted to release something but we can't do that, we can't be reckless," stated Thomas Matsuda, head of the agency (Hofschneider 2015a). According to Earthjustice attorney Paul Achitoff, this explanation was "nonsense" (Hofschneider 2015a). Following the media coverage, the HDOA eventually released to the public minimal data of total restricted-use pesticide (RUP) sales records not specified by company or user. This is data they have readily available on file and could provide to the public (Hooser 2015c).

At the same time as the state has refused to regulate pesticide use, it has also worked to facilitate companies' exemptions from federal pesticide laws (Pala 2015; Hooser 2015d). The Hawai'i Department of Agriculture has granted exemptions on federal pesticide label requirements, including allowing pesticides to be used in higher wind speed conditions and increasing the frequency of allowable applications (HDOA 2015). The state's Agribusiness Development Corporation has facilitated exemption from compliance with Clean Water Act standards by extending a decades-old sugar plantation permit to discharge toxic chemicals into the water (National Pollutant Discharge Elimination System permit). The exemption was recommended by the ADC because the Clean Water Act criteria would "likely be extremely difficult to meet" with more stringent federal limits for pesticide discharge now in force (JFFG Draft 2016, 46–47). Following litigation by Earthjustice and community groups, in 2019 a federal judge ruled that the ADC has been violating the Clean Water Act by allowing the discharge of millions of gallons of water contaminated with pesticides, sediment, and heavy metals into the ocean each day (Cruz 2019). Strong parallels can be drawn to the way the state advocated and assisted exemption of sugar plantations from federal environmental laws passed in the 1970s (Kahane and Mardfin 1987). Exemptions from county laws regarding grading and grubbing of soils are also granted to companies.

Non-enforcement of existing law also enables externalization of industry health and pollution costs. The Hawai'i Department of Agriculture is still reviewing pesticide inspection reports from several years back. It has not followed up on a vast majority of complaints of potential pesticide law violation (Hofschneider 2014b; Cocke 2013c). On the island of Kaua'i, numerous ongoing violations of county environmental and conservation law have been revealed, with one council member alleging that this "has been the norm" (D'Angelo 2014). In sum, referring also to pineapple and sugar plantation days, the Hawai'i Center for Food Safety declares: "Hawai'i State officials have a history of covering up pesticide contamination and denying clear health risks to citizens in order to protect agricultural interests" (Freese, Lukens, and Anjomshoaa 2015, 18).

Due to powerful and persistent social movement pressure, in 2018 the Hawaii State Legislature did finally pass the first meaningful regulations related to agrochemical-seed-biotech operations. Senate Bill 3095 bans the use of the highly dangerous neurotoxin chlorpyrifos, mandates 100-foot no-spray buffer zones around schools during school hours for RUPs, and requires annual reporting of RUP usage. The law was passed following the Trump administration's reversal of an Obama administration order to ban Dow Chemical's (now Corteva's) chlorpyrifos in agriculture. Chlorpyrifos was already banned for home use nearly twenty years ago due to its harmful impacts on the developing brains of children and fetuses. The Hawai'i ban was widely hailed as setting precedent for other states to follow in lieu of federal inaction.

While Hawai'i's 2018 chlorpyrifos ban and disclosure bill was an important win, the state's policy framework for ensuring that the public does not bear the health and environmental impacts of agrochemical operations remains woefully inadequate. Hundred-foot pesticide buffer zones are scientifically proven to be insufficient for protecting children from exposure. Annual pesticide reporting falls far short of what is necessary to know if exposure has occurred, conduct health and environmental studies, and protect communities living near pesticide-intensive agriculture. Moreover, the reporting has been exceptionally convoluted and difficult to make sense of. The fight for SB3095 illustrates the extent to which the state has resisted regulating big agribusiness and how much social movement pressure it took just to begin to force some environmental and health protections.

In addition to pesticides, there are a range of environmental impacts associated with genetically engineered crops, including concerns specific to Hawai'i's unique biodiversity and ecology. The Hawaii State Legislature has enacted only one law regulating GMOs, in 1988, which has never been amended. The single law requires merely that a copy of federal field testing permits be submitted to the Department of Health. The state is permitted to ask for redacted confidential business information, but it chooses not to. This may be an intentional decision so as to avoid having to disclose information about the industry to the public (Gibson 2014, 242). A recent activist inquiry into these records revealed that notifications were likely never examined by state regulators and that most of the information was redacted. The state does not require disclosure of the location of experimental GMO field trials, irrespective of their nature and potential impacts. Bills to require further industry disclosure have been introduced but not passed. In the words of Earthjustice attorney Paul

Achitoff, the state merely acts as a "rubber stamp" regarding any federal decisions about GMO crop trials (Gibson 2014, 242).

Local Power Webs

As examination of land, water, institutional, tax and policy subsidies to agrochemical-seed-biotech operations shows, the industry relies significantly on local collaborators and government to materialize its interests. Local alliances are essential not only for shaping policy and ensuring access to resources, but also for influencing public opinion and marginalizing dissent. Most important, a wider sociopolitical landscape that is favorable to the industry is constantly under reproduction by Hawai'i's most powerful. The agrochemical industry largely stepped into and has benefited from existing arrangements. At the same time, their wealth and influence shape new allegiances. In this sense, already existing local power conditions smooth the industry's operations, while the industry also brings outsized political and financial influence to bear upon Hawai'i and its local ruling circles.

Decades after King Sugar, concentration of political and economic power in the islands remains remarkable. Power operates especially through interlocking networks and institutions and is significantly tied to land control. Previous Senate Majority Leader Gary Hooser writes that "in addition to 'running the plantation,' it is no secret that the large landowners have historically also 'ran the government.' " From his own insider perspective, he argues that while political parties flipped in the Democratic Revolution, the underlying power structure remains largely the same (2017). Noel Kent elaborates that those with the most power in Hawai'i typically remain loyal—and tied through multiple networks—to the local financial establishment, leaders of chief industries, developers, major landowners, and the dominant unions (Kent 1993, 1994). Similarly, historian Robert Stauffer contends that in no other US state has a "political machine" ruled for so long (2001, 101). Moreover, he suggests that "public boards and commissions take on a membership that reflects the machine more than in other states" (Stauffer 2001, 101). Other Hawai'i scholars identify a "quasi-democratic variant" of a "top-down, centralized, plantation era political system" (Rohter 2001, 2), the "ruling circles in the state" (Aoudé 2001, xx), "politics and business in the Islands continue an intricate cross-pollination" (HawaiiHistory.org), or more colloquially, the "good-old-boy system."

One place these trends and alliances make themselves visible every legislative year is in the lobbying positions taken on state bills. For example, in 2015 a bill widely supported by the general public for pesticide buffer zones around schools and hospitals (HB1514) was heavily opposed and successfully defeated by the agrochemical-seed-biotech industry, the Hawai'i Department of Agriculture, the Hawai'i Farm Bureau, the Hawaii Cattlemen's Council, the Chamber of Commerce Hawaii, the Hawaii Agriculture Research Center, large landowners, the Hawai'i Coffee Growers Association, and individual members of University of Hawai'i's CTAHR and College of Agriculture, Forestry and Natural Resource Management (Hilo campus). The same year, a "Right to Farm" bill drafted by and for the industry with the goal of stripping counties' regulatory rights was supported by the Hawai'i Department of Agriculture, the Hawai'i Farm Bureau, large landowners, the remaining sugar plantation, the International Longshore and Warehouse Union, the Chamber of Commerce Hawaii, and a lobby group representing landowners. While the vast majority of testimony from the public was in opposition to what was dubbed the "Hawai'i Monsanto Protection Act," the small coalition of its supporters reflects some of the most powerful institutions and private interests in Hawai'i. The power network that supports the agrochemical industry is active through each legislative session and beyond, and this has come to be expected by legislators and activists whenever a bill related to the industry is under consideration.**

Among the institutions that serve to maintain the material interests of the agrochemical industry, the Hawai'i Farm Bureau is most notable. It brands itself the "Voice of Agriculture" and is largely regarded as such. It is part of the American Farm Bureau Federation, which was founded in 1919 to counteract farmer and labor organizing that was becoming a strong leftist force (Hauter 2012). Since then, it has played a monumental role for over a century advancing the interests of dominant agribusiness corporations over farmworkers, farmers, consumers, and the general public. Side-by-side with the Farm Bureau, the Hawaii Cattlemen's Council similarly advocates the interests of dominant agribusiness and is unified with the National Cattlemen's Beef Association, the main lobbying group for the US cattle industry.

Like the Farm Bureau, the Chamber of Commerce Hawaii hails itself the "Voice of Business" in ways that intentionally belie *which* business interests it actually advances. Founded two years after the Māhele by ten white businessmen, the Hawai'i Chamber has a history of pushing the interests of the islands' elites for more than 150 years. Today, its "gold sponsors" and board of directors on the island of Kaua'i include Dow, DuPont, and Syngenta. The Hawai'i Chamber is an extension of the US Chamber of Commerce, the largest lobbying group in the US. The Chamber is most prominently used as a proxy

** For further examples, see HB673 and SB590 in the 2013 legislative session and HB 790 in the 2017 legislative session.

by agribusiness, pharmaceutical, tobacco, oil, and other corporations through which to "pursue their less popular causes anonymously" ("The Chamber of Secrets" 2012; "Alyssa Katz on the Influence Machine" 2015; Hauter 2012).

The Hawai'i Agriculture Research Center is a rebranding of the Hawaiian Sugar Planters' Association, which carried out research and coordinated everything from policy advocacy to wage control for the plantation oligarchs (MacLennan 2014). In 1996 the Hawaiian Sugar Planters' Association changed its name and expanded its research to include an emphasis on genetic engineering. The Hawai'i Agriculture Research Center runs education and outreach programs that are reflective of its history supporting dominant plantation interests in the islands, promoting monocropped industrial-style agriculture, and dismissing scientific concerns around pesticides (Valenzuela 2012, 12). Its board of directors includes many overlapping Farm Bureau directors, as well as corporate landowner managers and a previous director of the Hawai'i Department of Agriculture.

Also consistently championing the interests of the agrochemical-seed-biotech industry are nonprofit economic development boards, which were formed in the early 1980s on the islands of Kaua'i, O'ahu, Hawai'i, and Maui/Moloka'i. Board membership reflects how power is maintained through social networks. The Kauai Economic Development Board (KEDB) executive committee is made up of representatives from the largest corporate landowners, managers from DuPont and Dow, a past president of the Kauai County Farm Bureau, the director of the county's Office of Economic Development, the previous county attorney who now heads a law firm that primarily deals with real estate, the president of the island's coffee plantation (which leases land to agrochemical companies), the head of the Kauai Visitors Bureau, two military contractor companies, a commercial real estate manager, and the chancellor of Kaua'i Community College. The KEDB was contracted by the county to develop the official Kauai Economic Development Plan, in which the political positions and investments of dominant economic interests, including the agrochemical industry, are clearly reflected. The KEDB also manages the government-sponsored West Kauai Technology and Visitor Center, which hosts only agrochemical industry and military contractors.

Revolving roles within circles of power are exceptionally common. For instance, with the introduction of Kaua'i's pesticide disclosure bill (2491), the industry rapidly hired the previous governor's liaison, the CEO of the Kauai Economic Development Board, and the previous general manager of the largest hotel for community outreach. In particular, there is an intimate relationship and cross-pollination between regulators and the regulated. Following Bill 2491, the mayor's top administrator was hired by Syngenta. She was then appointed by the mayor to the Kaua'i Board of Water Supply, which sets policy related to county water use. At the state level, one senator describes it an "open secret" that key power holders, historically including the chairs of committees that control legislation related to agriculture, have a close relationship with the agrochemical industry (Pala 2015). This includes many legislators receiving tens of thousands of dollars annually just in direct campaign contributions from the industry (Grube 2014). Previous governor Neil Abercrombie, himself a primary beneficiary of industry money, summarized this cozy relationship in his proclamation to the industry: "I'm here lobbying you . . . You don't have to lobby me. You don't have to lobby Russell [Department of Agriculture chair]. You don't have to lobby this administration!"

Though revolving doors, payoffs to politicians, and entrenched institutions working on behalf of the powerful are common, none of this is without nuance and contradiction. The local institutions and actors that facilitate agrochemical industry occupation of the islands do not signify a monolithic and uncontested power bloc or even uniformity of interests. There is persistent and growing contestation within institutions like the Farm Bureau, Cattlemen's Association, and Economic Development Boards that indicate internal dissent around support for the agrochemical industry. Regulators themselves have dramatically changed their positions on regulating the agrochemical industry as social movement pressure grows. Though there is no singular conspiracy of local power or unchallengeable ruling order, there are clearly identifiable dominant institutions, actors, pathways, and mentalities that are shaped by structural power and wealth. This structural wealth and power, as well as its ideological pillars, are inseparable from the functioning of US imperialism and the ongoing American occupation of Hawai'i.

United States Regulatory Regimes

While robust support from the local state and power holders is critical to the agrochemical-seed-biotech industry's decision to locate in the islands, the local state itself is most fundamentally defined and confined by American occupation and subjection to United States rule. It is Hawai'i's place both within and outside of the US that is most important to the industry's decision to locate in the islands. The US is the primary adopter and promoter of agricultural biotechnology and provides the largest GMO seed market. Operating within US territory enables easy transfer through phases of seed development and distribution, as well as between research and commercialization. In addition to fluidity of production, the federal government provides conditions without which the industry could not function. As discussed earlier, strong states are vital to securing property rights, capitalist markets, administrative order, and bureaucracy. In the case of the agrochemical-seed-biotech oligopoly, these factors are most remarkable in the strict patent, trade secret, and other property protections provided by the American state. The US has worked hand in hand with the industry to extend its products and property rights globally, to privatize publicly funded science and technology in the interests of the industry, and to set the conditions for it to morph into a highly consolidated oligopoly.

While it is under dominion of US law, Hawai'i is subtropical and distant from the continental United States. Unlike the rest of the country, it has a year-round growing season that enables much faster development of agrochemical-seed-biotech products. Furthermore, Hawai'i is isolated from other agriculture that could contaminate or be contaminated by experimental operations. As explained by University of Hawai'i plant breeder James Brewbaker, "If you're in the middle of Iowa and studying a new [biotech crop], you've got to be in absolute isolation. Here it is concentrated in a way so that you'd think about all of [their acres are] made available for genetically modified crops" (Voosen 2011). This isolation also means that lucrative corn seed is less exposed to pests and disease that plague monocrop production in the Corn Belt. If a virus like corn rust is spotted, production in the islands can shift to another area that has not been intensively farmed in corn, a strategy that is not available in the Midwest United States (Schrager 2014). While geographically isolated and subtropical, the islands are at the same time bound by US regulatory regimes. It is this "inside" and "outside" relationship that also makes the island colony of Puerto Rico a main site of GMO crop field trials; it historically has the second highest number of test sites behind Hawai'i and has recently surpassed it (USDA 2020).

Speaking to the *New York Times* about its operations in Hawai'i, a spokeswoman for DuPont Pioneer stated simply, "We like being able to work under the U.S. regulatory network" (Voosen 2011). In addition to monopoly intellectual property rights and fluidity between phases of production, the appeal of the US regulatory network is the largely anti-regulatory approach taken to pesticides and GMOs. US policy related to genetically engineered crops originated in the doctrine that "the market, rather than the law, was the right instrument for controlling the inventiveness of biotech" (Jasanoff 2005, 63). The GMO crop experimentation and seed development conducted in Hawai'i falls under the federal Coordinated Framework for the Regulation of Biotechnology. The Coordinated Framework is a patchwork of regulations with significant gaps and inconsistencies, which was adopted from a Monsanto written proposal in the 1980s (ABSA 2015; see chapter 2).

Under the Coordinated Framework, the USDA is tasked with oversight of

GMO testing and product deregulation. It is also in charge of promoting US agricultural interests, a mandate that frequently collides with its regulatory role. The USDA requires only 1 percent of all GMO field trials to go through an environmental impact statement or environmental assessment. The remaining 99 percent go through a notification process where risks such as harms to human health or the environment are not considered and companies perform their own risk evaluations (Gibson 2014). The last full environmental assessment of a field trial in Hawai'i was conducted in 1994 (ISB 2015). Companies are able to conceal the most basic information about their experimental field trials as trade secrets, including where they are located, what they are testing, and how they are using pesticides. While any state government has the legal authority to review a GMO developer's federal permit application, it cannot block a field trial from occurring. Once up and running, the USDA does not inspect all field trial sites, instead using a "risk-based approach" to select sites for inspection (USGAO 2008, 18). When inspections do take place, they are limited to tasks like ensuring hygiene, verifying isolation distance of crops, and reviewing planting and harvesting records (Gibson 2014, 238). The USDA's very partial inspection approach has led to escape of unapproved GMO material.

After a genetically engineered crop has gone through the testing phase, the USDA decides whether a company must prepare an environmental impact statement prior to deregulation and commercialization. Judgment is based on potential to "significantly affect the quality of the environment" (40 CFR §§ 1508.9[a]). Of the ninety crops deregulated as of 2012, two environmental impact statements were conducted as a result of lawsuits and court orders (Montgomery 2012). Once deregulated, genetically engineered crops are grown with no more oversight than any other crop. As summarized by a US Government Accountability Office report, there is no "program for monitoring the use of marketed GE crops to determine whether the spread of genetic traits is causing undesirable effects on the environment, non-GE segments of agriculture, or food safety, as recommended by the National Research Council and others" (USGAO 2008).

The EPA has primary authority over pesticide regulation. It does not have a single inspector in Hawai'i and delegates local pesticide monitoring duties to the Hawai'i Department of Agriculture. In its role, the Hawai'i Department of Agriculture investigated seven of seventy-two possible pesticide violations on one island alone in 2011–2012. On average these took several years for follow-up (Cocke 2013c). According to the employee assigned to review pesticide violations, there has been "little if any action against pesticide misuse" (Cocke 2013c). The public is not notified of violations, even when these might have impacted surrounding communities. Pesticide drift is almost entirely unmonitored, and the labeling law likely broken regularly in regard to spraying in Hawai'i's frequently windy conditions. The EPA fails to take into account the reality that some populations, such as farmworkers and those living in agricultural production areas, are more vulnerable to daily pesticide exposure. There is no program in place for regular testing of pesticide contamination in the soil, air, or water.

The EPA determines which pesticides can be used in the United States and how they can be used. Unlike many other countries that adhere to a more precautionary approach, the EPA regulates pesticides according to a risk-benefit standard that weighs purported economic benefits against proven or potential harms. Thus, the EPA's allowable levels of use and safety standards for air, water, soil, and food residue are typically far less stringent than those of European and other countries. The biases of the US regulatory system are highlighted by the fact that foreign agrochemical-seed-biotech corporations use pesticides like atrazine, alachlor, and permethrin in Hawai'i that are banned in their home countries. Notable features of the EPA's regulatory process for pesticides include that tests are conducted only on single active ingredients in a formula despite the fact that some so-called inert ingredients may be toxic and synergistic (i.e., a pesticide is not actually tested in its whole formula), tests are conducted on one pesticide at a time and not in their synergistic effects as they are experienced in the real world, testing is biased toward acute effects and largely neglects impacts from long-term lower-level exposures, there is a near exclusive reliance on animal experiments conducted by chemical companies themselves, little attention is paid to more relevant human epidemiological studies carried out by independent scientists, inconsistent definitions of "reasonable risk" are used, and label instructions assume perfect compliance with a range of complicated directions that are often unworkable and not followed (Freese, Lukens, and Anjomshoaa 2015; Sass and Wu, n.d.).

The US policy regime governing pesticides and agricultural biotechnology allows the industry to conduct their Hawai'i operations in such ways that generate substantial public costs. These are significant subsidies—largely unknown—that the public will be paying for long after the agrochemical-seed-biotech industry has exited the islands. The public is still paying the costs of sugar and pineapple plantations past, with Superfund sites, contaminated drinking water systems, and overall "unusually widespread occupational and general population exposures" to pesticides (Allen, Gottlieb, et al. 1997, 679; Cocke 2013a).

In many ways, today's plantations run by Bayer-Monsanto and DowDuPont do not stray far from those of the sugar oligarchy. Much of sugar's infrastructures, institutions, and ideological underpinnings have been directly inherited. Agrochemical-seed-biotech occupations similarly operate locally by way of consolidated power and resource control, undergirded by US occupation and its military and economic interests. As with plantation sugar, benefits are privatized and costs are socialized, with disproportionate impacts across race and class. While for decades virtually every local politician has dutifully spoken to the need for "economic diversification," what is almost entirely disregarded is the question of what it would mean to actually democratize and distribute Hawai'i's abundant wealth and topple deep structural inequalities.

To consider actual alternatives is to look to the very root conditions that continue to facilitate plantations, oligarchies, and oligopolies. It is to seek not a mere diversification of Hawai'i's highly unequal and antidemocratic monoeconomy—from a little less Hilton tourism to a little more Monsanto agriculture—but to disrupt the very systems and powers that preclude truly alternative possibilities. Without a doubt, material and ideological conditions of the present—including compulsions emanating from beyond the islands' shores—restrict what can be thought and done. But in opposition to the alibis of injustice and to the incessant realisms that cast the current order as inevitable, justice demands claiming those very material limitations as precisely the reason that deep systemic change is most necessary. Resistance and the fight over the future is the subject to which the remainder of this book now turns.

Chapter 7

Resistance Is Fertile

Two years after Kaua'i's historic Mana March flooded the streets with residents demanding the passage of Bill 2491 for pesticide disclosure and buffer zones, an estimated ten thousand people joined in the Aloha 'Āina Unity March through Waikīkī. The event inspired mass participation following a federal court overturn of laws in three counties regulating the agrochemical-seed-biotech industry. Reflective of growing solidarity and mobilization around interconnected issues, the march united concerns over pesticides and GMOs, land use and development, protection of Mauna Kea, militarization, and the rights of Native Hawaiians. A few years later, the movement succeeded in passing the country's first ever ban on the highly dangerous neurotoxin chlorpyrifos, used widely by the agrochemical-seed-biotech industry in Hawai'i. While suffering some major legislative and legal defeats, the movement has grown in size, breadth, and diversity. Most important, it has been central to the development of intersectional movements across the islands. The growing wave of progressive movements in Hawai'i is both reflective of global trends and effecting global trends. As Native Hawaiian activist Walter Ritte says, "These five major chemical companies chose us to be their center. So whatever we do is going to impact everybody in the world" (Altemus-Williams 2013).

The Eruption of Politics in Bill 2491

Activists dub Hawai'i as GMO Ground Zero—geographically isolated islands where the battle over the power and injustices of the agrochemical-seed-biotech oligopoly has been fierce. While perhaps limited in its centering of a single technology, the description calls attention to the islands as an epicenter of industry research and development, including the use of genetic engineering to create plants that are resistant to pesticides. Political organizing around GMOs first took root in the islands in the early 2000s. Activists focused especially on corporate ownership of seed and strategies of seed saving and exchange. A strong wave of resistance led by Kānaka Maoli emerged in 2005 to the patenting and genetic engineering of their traditional staple food crop kalo (taro). Kalo is considered by Hawaiians to be their sacred elder brother and the staff of life. Walter Ritte likened the private ownership and genetic manipulation of kalo to the original enclosure of the commons, now robbing spiritual power: "Biotechnology is the second Mahele . . . the Mana Māhele" (Trask 2006, 24). Following tremendous pressure and eventually direct action by Kānaka Maoli to chain the doors of a University of Hawaiʻi building, the university dropped and publicly destroyed its patents on three varieties of taro in 2006. Two years later, the first ordinance related to GMOs was passed by the County of Hawaiʻi (which encompasses Hawaiʻi Island), banning genetically engineered taro and coffee on that island.

Connected to and partly growing out of these movements, since 2013 especially there has been an eruption of politics around the local impacts and implications of the agrochemical-seed-biotech industry's occupation of the islands. At the opening of the 2013 state legislative session, hundreds gathered at the capitol for the Idle No More: We the People rally. Mounting concerns over the agrochemical-seed-biotech industry were central. After a tour through the islands, Indian activist Vandana Shiva proclaimed at the state capitol that land, biodiversity, and cultural heritage must displace Monsanto from the center of the economy. Hawaiian activist Andre Perez concluded, "The things that we're standing up against are really at the core of capitalism" (Altemus-Williams 2013). A few months later, thousands took to the streets in five spirited marches across the main Hawaiian islands. "Aʻole GMO" and "Aloha ʻĀina" animated chants and signs. Aloha ʻāina is most commonly defined today as "love of the land." However, aloha ʻāina is mutually a political concept with a long history in Hawaiian anticolonial and independence struggle (Silva 2004). It is a concept that encompasses spiritual belief and the reciprocal relationship between people and ʻāina, as well as Native Hawaiian rights to sovereignty and protection of land (McGregor 2007, 265). Many activists intentionally invoke aloha ʻāina to convey that their struggle is not merely over a single technology or corporation, but more systemically about food, land, sovereignty, and an entirely different worldview (see chapter 9).

The growth of action in the streets beginning in 2013 corresponded with major land expansion and more than 500 percent "value growth" in the industry since 2000 (Loudat and Kasturi 2013). The spread of the industry has been accompanied by school poisonings, extreme dust blowing into people's homes, and suspicion of health and environmental impacts from pesticides (see chapter 3). While many activists are broad in their criticisms of the industry, it is the issue of local pesticide use that has most galvanized widespread public attention and legislative action. Following marches and protests in early 2013, in June the first major county law aiming to regulate the agrochemical industry's pesticide use was proposed on Kaua'i. Prior to Bill 2491, several events on Kaua'i drew broad public concern. Most notable were the series of school poisonings at Waimea Canyon Middle School and a class action lawsuit over the impacts of excessive fugitive dust and pesticides. Residents had raised concerns over high levels of dust blanketing their homes since 2000 and finally turned to litigation in 2011. A group of more than 150 residents from the small town of Waimea filed a state court lawsuit against Pioneer DuPont and Robinson Family Partners, which leases land to DuPont. Gary Hooser, the council member who co-introduced Bill 2491 with Tim Bynum, noted that the Waimea class action lawsuit indicated widespread and deep public concern: "for local residents to sue the primary employer in a town like Waimea is unheard of. It really showed me that this issue has cut deep into the community" (personal communication, 2013).

Gary Hooser had previously spent many years as a state senator, where he introduced multiple pesticide-related bills following the Waimea Canyon Middle School poisonings from 2006 to 2008. During Hooser's time, pesticide-related bills rarely made it to the first step of being scheduled for a hearing and none passed into law. Hooser and activists heavily criticized the state for "dereliction of duty," "incompetence," and catering to industry interests (Hooser 2015b). It was largely the inability to gain traction at the state level that motivated Hooser and others to attempt a county initiative in 2013.

Bill 2491 began with a living room gathering of several friends in their twenties and thirties who asked Hooser to come meet with them and posted an open social media invitation. Most of these young activists were born and raised on the island and had never been involved in politics. Following the initial small meeting, the group spent months conducting extensive research, building a decentralized coalition of core organizers, and reaching out to organizations with scientific and legal expertise, including Earthjustice, the Center for Food Safety, and Pesticide Action Network. Though they sought council from national organizations, the local people working on the bill were not paid by or associated with an NGO. According to Hooser, a longtime politician, the process was more engaged than any he had previously witnessed: "it wasn't just activists showing up, it was activists really doing their homework, like learning to read attorney opinions and pesticide Material Safety Data Sheets. They really took ownership" (pers. comm., 2013). For many, the experience was a rapid political education that broadened to much wider systemic concerns and initiated a longer commitment to activism. Hooser frequently describes the process of Bill 2491 as "true grassroots democracy in action."

While the local motivations for and design of Bill 2491 centered on pesticides, the concerns of core organizers also reflected wider understandings of food system injustices. One young participant who was involved early in Bill 2491 articulated her repudiation: "Multinational companies [are] seeking profit in our food system. Some of them are chemical companies, and they serve no useful purpose on this planet except for making profit for their shareholders. In Hawai'i we have no business hosting these corporations who want to do their experimentation and patent their seeds" (pers. comm., 2013).

Another initiator of Bill 2491 similarly described that many who were involved had been "watching everything happening globally around Monsanto, and see it on the local level, and see their place in it in the bigger picture" (pers. comm., 2013). For others, it was local involvement that broadened their awareness. Some expressed openness to GMO technology, but that "nobody can dispute that what is happening here [in Hawai'i] is not a good thing" (pers. comm., 2013). At the same time as activists asserted their opposition to local and global impacts of the industry, they also articulated that they were fighting for an alternative future: "This is about what we want for our future, what agriculture means to us, what happens to our land in the long run, where we get our food and if our agricultural lands are capable of food production into the future. This is about what we create moving forward" (Fern Anuenue, Facebook post, 2015).

After months of research and organizing, Bill 2491 was introduced in June 2013. Dubbed Kaua'i's Right to Know Bill, in its original form it contained mandatory disclosure of all pesticide use including date and time, location, and quantity; warning signage prior to pesticide spraying; annual disclosure of GMOs being grown; 500-foot pesticide buffer zones around schools, hospitals, residential areas, public roads, shorelines, and waterways; a prohibition on open-air testing of experimental pesticides; a temporary moratorium on expansion of GMO operations until an environmental impact study was completed; and a mandate for the county to complete an environmental impact study on the impacts of agrochemical-GMO operations on the island. The pesticide elements of the bill were to pertain to commercial agricultural entities using or purchasing more than five pounds or fifteen gallons of restricted-use pesticides annually. This applied to only Dow, DuPont, Syngenta, BASF, and Kauai Coffee (the largest coffee plantation in the US). Monsanto and Bayer were not present on Kaua'i, though they were on other islands.

In September, Bill 2491 was significantly amended to exclude the temporary moratorium on expansion of agrochemical-GMO operations and the ban on open-air testing of experimental pesticides. Pesticide buffer zones were reduced to 100 feet around public roads, shorelines, and waterways, and the environmental impact study portion of the bill was changed to a community-based process described as an environmental and public health impact study. In October, the bill passed after a nineteen-hour meeting that included multiple dramatic attempts at deferral and derailment. Mayor Bernard Carvalho vetoed the bill a couple of weeks later. He simultaneously recruited one of the council members who voted for the bill to his administration, leaving the council short of the votes needed for a veto override. The remaining supporters of the bill were able to select a new council member with just one day left for a veto override, and the bill finally passed into law without the mayor's signature on November 16, 2013. In its final form, Bill 2491 included pesticide disclosure, buffer zones, and an environmental and health impact study.

Bill 2491's passage set national precedent as the first local government law in the United States to regulate agrochemical companies at the site of GMO seed cultivation and field trials. It was the most substantial regulation of corporate agribusiness ever passed at the county level in Hawai'i, until it was followed in 2014 by a bolder citizens' initiative on Maui. Kaua'i was described in formal and social media a "true David versus Goliath," "the little engine that could," and "the hotbed of anti-GMO activism." At the time of its passage, the bill was widely heralded by national activists as a tipping point in the struggle against the agrochemical-seed-biotech corporations' influence over the food system. The political uprising that unfolded around Bill 2491 was unfunded, almost entirely outside of formal nonprofits, and highly decentralized, though core organizers collectively carried many integral tasks forward. While small amounts of donation money trickled in to more formalized groups for things like printing flyers, it was the countless hours of unpaid labor by grassroots activists that pushed the David versus Goliath struggle forward.

Over its nearly five-month course, Bill 2491 generated unparalleled public testimony and controversy. Activists, workers, and council members sat through hundreds of hours of public hearings and meetings. Lines to get into council meetings sometimes started the afternoon before meetings began and involved multiple campouts in rain and thunder. In the largest public demonstration of the island's history, an estimated two to four thousand people marched in support of Bill 2491. Chemical companies also organized smaller sign-waving demonstrations with dozens of people. Weekly headlines, spotlights in national news, daily letters to the editor, continual public radio commentary, and a general buzz around the bill was felt even by those who were not involved.

The bill created a space for new voices, information, and political organizing to emerge. One mother living adjacent to chemical company fields said, "I couldn't talk about this stuff in my community because they fund everything and they provide jobs. I had no idea that so many other people in my community felt this way until we took to the streets" (pers. comm., 2013). Increasingly, directly impacted residents told their stories of experience with the industry. Information began to surface about workers' grievances, illegal spraying practices, violations of tax and land use laws, and strange environmental and health occurrences. While much in these resident reports was anecdotal, the allegations pointed to the urgent need for further study. Most significantly, more than fifty local doctors, nurses, and mental health providers formally submitted concerns to the county council, with many referencing their direct observations of "a lot of major health anomalies" in communities living near fields (Stop Poisoning Paradise 2013).

National and global recognition of Hawai'i's place in the agrochemical industry's chains of production was spotlighted during the campaign for Kaua'i's Bill 2491. Initially, this was largely achieved through activists' persistence and strategy. Organizers generated frequent press releases, used social media extensively, crafted their own articles for the media, and built relationships with news media, especially with independent news outlets. Their success in reaching progressive independent media persuaded more mainstream outlets to pay attention. Significant, in-depth pieces were featured in the *New York Times, The Guardian, Al Jazeera, PBS*, the *Washington Post*, and many other outlets, with the passage of Bill 2491 circling through the corporate news headlines. This media attention has not always or necessarily been helpful to impacted residents, activists, or the goals of justice. However, it has been one element in bringing the issue further to light and connecting to wider related struggles.

Though blocked by the courts and corporate power, the continuing significance of Bill 2491 in galvanizing collective action around agrochemical industry operations in Hawai'i cannot be understated. It was a point of convergence and unity, and a springboard for organizing that followed. Its passage marked a moment of collective hope for both local organizers and worldwide onlookers. As described by an initiator of the bill: "It was a seminal moment and a catalyst. It's the pebble in the water thing. Some people were touched in a way they will never forget. We need more moments like these, we have to win more, because people are touched and they believe when they experience that. I'm sure those people that were part of winning will tell their grandkids" (pers. comm., 2013).

Islands Rising

The ripples of Bill 2491 spread across the island chain, converging with already energetic movements on other islands and bolstered by interisland organizing. At the same time as Kaua'i's Right to Know Bill was playing out, the County of Hawai'i introduced its own bill that would prevent the agrochemical-seed-biotech companies from ever establishing operations there. Currently, none of the agrochemical companies operate on Hawai'i Island. Hawai'i Island's Bill 113 prohibited open-air cultivation, propagation, development, and testing of genetically engineered plants other than GMO papaya, which is already widely grown. The bill exempted cultivation and testing of GMOs in enclosed areas. Just days after Kaua'i, the Hawai'i County Council passed Bill 113 by a vote of 6–3. In his signing of the bill into law, Hawai'i Island mayor Billy Kenoi stated: "Our community has a deep connection and respect for our land, and we all understand we must protect our island and preserve our precious natural resources . . . With this new ordinance we are conveying that instead of global agribusiness corporations, we want to encourage and support community-based farming and ranching" (2013). Kenoi's carefully chosen words were a testament not to his personal political record but to shifts in public discourse resulting from the movement.

Most remarkable, one year later voters in Maui County (including the islands of Moloka'i and Lāna'i) passed a ballot initiative placing a temporary moratorium on all genetically engineered crop production and experimentation pending an environmental and health impact study. With Monsanto and Dow occupying around 3,500 acres on Maui and Moloka'i at the time of the ballot initiative, its passage was the strongest local regulation of the industry in the nation. Activists gathered 9,062 valid signatures to get the first-ever citizens' initiative on the ballot. The initiative passed with 50 percent of the vote, despite the industry outspending advocates by at least 100–1. Monsanto and Dow spent nearly $8 million just in reported direct funding against the initiative—more than any group on any election in Hawai'i ever—a price of more than $362 for every no vote (Kerr 2014). Advocates raised less than $90,000, primarily through crowdfunding and donations averaging $50 ("Monsanto & Dow Spend More than $8 Million" 2014). It was dubbed the "Maui Miracle," and the Center for Public Integrity called it the most expensive local initiative in the country that year. Though Maui voters supported the initiative, Maui County officials opposed it and were sued by its drafters days after passage to require the law's enforcement.

The passage of bold regulations of the agrochemical-seed-biotech industry by three of four island counties—one by citizens' initiative—was momentous.

It directly challenged the industry's global production lines and their future in the islands. It also set a model for local government to regulate where federal and state government has refused to. Alarmed by the precedent-setting significance of the new laws, the industry promptly sued to block all three of them. Using courts to block regulation, claim monopolistic property rights, and augment oligopolistic power is a well-rehearsed and heavily funded strategy of the agrochemical industry. On November 18, 2016, the United States Court of Appeals for the Ninth Circuit invalidated all three county laws based on federal and/or state preemption. Preemption refers to laws at higher levels of government that take precedence over those at lower levels. The Ninth Circuit Court ruling laid the authority for pesticide regulation on the state and the authority for GMO crop regulation on the federal government. The court noted the lack of any express preemption of the statutes or any direct conflict with existing state statutes. Instead, it determined that the local ordinances were *impliedly* preempted based on the existence of regulatory schemes already addressing pesticides and GMOs. Hawai'i does not expressly preempt local authority from regulating pesticides or GMOs, as some other states do (a result of a concerted campaign in the 1980s by the chemical industry to strip the control of local municipalities over pesticide regulation). The Ninth Circuit's decision to invalidate local pesticide and GMO regulations on implied preemption grounds was the first of its kind, and according to Chapman University professor of law Rita Barnett-Rose, "will likely have a significant adverse impact on the ability of other local governments throughout the U.S. to regulate" (Barnett-Rose 2015, 73). If they stand, the decisions may also impact local government authority to regulate on other issues of local concern.

The Ninth Circuit also called the question of Hawai'i's constitutional public trust obligations "irrelevant." Ten years earlier the Hawai'i Supreme Court ruled that the public trust duties of the state also apply to counties, including preserving and protecting natural resources. David Forman, director of the Environmental Law Program at the University of Hawai'i at Mānoa's William S. Richardson School of Law, argues that the Supreme Court's earlier decision means that the public trust obligations of counties "are self-executing; they do not lie dormant awaiting express state law provisions granting legislative power to the counties" (Forman, n.d., 14). He writes, "The federal courts' flawed and dismissive treatment of Hawai'i constitutional provisions that reflect the intergenerational equity/precautionary framework, utterly failed to address numerous gaps in federal and state laws regulating the GE/GMO seed industry in Hawai'i" (Forman, n.d., 14). Many other legal experts have similarly argued that the Ninth Circuit's decision to invalidate the county laws was erroneous and not justified by existing federal or state law. It is widely believed that had the cases gone to the Hawai'i Supreme Court with its expertise in Hawai'i state law, they would have been decided in favor of the counties.

The Long, Slow Work of Change

Following the 2013–2014 historic passage and overturn of laws in three of four island counties, much attention has turned to regulatory action at the state level. The industry has focused most fiercely on laws that would explicitly preempt any county regulation of pesticides and agriculture. Typically framed in the language of "right to farm," a plethora of subtle and outright attempts were made each year to abolish county regulatory rights while the Ninth Circuit was still deliberating over county laws. On the other side, activists' primary focus has been on chlorpyrifos and passing state laws similar to Kaua'i's Bill 2491, including pesticide disclosure, buffer zones, and enhancing the pesticide monitoring and enforcement capacities of the Hawai'i Department of Agriculture. Other major legislative focuses of the movement have been on reforming or abolishing the Agribusiness Development Corporation, which leases public lands to the industry; incentives to support regenerative agroecological farming; pollinator protection; farm to school and local food procurement programs; subsidy for local food hubs; water rights and access; and banning the use of pesticides on school grounds.

For a few years following the major county actions, state bills that directly regulated the agrochemical industry got little traction. In 2017, twenty-five different bills aiming to regulate pesticide use were introduced. None passed. Key committees were chaired by known friends of the industry, and leadership in both the House and Senate was either similarly aligned or uninterested in taking the controversial issue on. The movement worked to build broader coalitional support, bolster their case with scientific evidence of the harms of pesticides, and challenge key gatekeepers and the balance of power in the legislature. Five years into intense legislative battles, in 2018—a major election year—the movement succeeded in passing a law banning chlorpyrifos, mandating 100-foot no-spray buffer zones around schools during school hours for restricted-use pesticides (RUPs), and requiring annual reporting of RUP usage (SB3095). While weak on disclosure and buffer zones, Hawai'i's first in the nation ban on chlorpyrifos—an extremely dangerous neurotoxin that has consistently been found in air and water sampling near fields—was a substantial win for local people and the environment. It also set precedent that other states are attempting to follow. Bans or restrictions have since been passed in New York, California, and Maryland and are in motion in multiple other states. A bipartisan bill for a national ban has also been proposed.

SB3095 succeeded in the end because of the long and persistent work of activists. They organized the combined voices of directly impacted residents and scientists who study chlorpyrifos. They built relationships with trade and labor organizations of teachers, nurses, and pediatricians that officially came out in support of the bill. They fought year after year, eventually wearing down the opposition and gaining champions of the cause in both the House and Senate. Gary Hooser reflected on his blog:

> [L]aws are not passed based on the cause being a righteous or urgent one. Unfortunately, they are also not passed just because a prestigious group of physicians provide incontrovertible evidence of a substance's harm. Passing a law of any significance requires a marshaling of forces to push and pull and cajole and threaten if needed (in a political election year sense), until the critical majority vote is achieved. There are always forces "on the other side", pushing back. The larger the stakes, the greater the push back and the tougher the fight. It is only through the marshaling of broad-based, strong, unrelenting community support that SB3095 was able to be passed through the legislature. (Hooser 2018)

Similarly, one year after the passage of SB3095, activist organization, coalition building, and persistence led to a ban of herbicide use at public schools. The ban was initiated by Hawai'i Department of Education superintendent Christina Kishimoto after activists requested a public meeting with the Board of Education and Dewayne Lee Johnson. Johnson, a Californian groundskeeper, is the first to win a major lawsuit against Monsanto over Roundup linked to his terminal cancer. Johnson traveled around the islands sharing his story with impacted residents, school groundskeepers, policymakers, and the wider public. His was one of many compelling speaking tours organized in an attempt to educate and push policy on pesticide use.

For both the industry and the resistance movement, a major focus for achieving legislative wins has been to change who is actually making laws. Following Kaua'i's Bill 2491, in 2014 several newly politicized young people entered mayoral, county council, and state representative races. None were successful, though they were part of major efforts to enroll new voters and set up new caucuses. They also significantly pushed the terrain of candidate debate on progressive issues. At the same time, an extensive, vicious, and well-funded effort by the industry succeeded in pushing Tim Bynum, co-introducer of 2491, off the council. The industry continued to defame Gary Hooser on a daily basis, including propagating widespread, slanderous lies. In 2016, Hooser also lost his seat on the council. The same year, the industry poured resources into electing a Syngenta employee to the council, who was later indicted for being the leader in a major methamphetamine drug trafficking ring. One of the first orders of business of the now heavily industry-influenced council was to repeal Ordinance 960 (Bill 2491). Given that the law had already been invalidated by the court, the repeal was largely a symbolic move.

While changes in the makeup of the Kaua'i County Council in 2014 and 2016 were a major hit to the movement, other counties and the state moved toward electing more supportive candidates. By 2018, strongly progressive candidates, including very visible movement leaders, took the majority of seats on Maui County Council. Others have been elected to the state legislature. Several newly elected policymakers across Hawai'i have been part of a candidate training effort called Kuleana Academy. The academy is designed and run by the Hawai'i Alliance for Progressive Action, which was "born of community struggles" on Kaua'i against the agrochemical-seed-biotech companies. Progressive social movement focus on elections and candidate training is necessarily broad, but issues around the agrochemical industry have been a key galvanizing force. As one longtime conservative politician complained, agrochemical activism is "changing the landscape of politics" (Eagle 2014).

The long and slow work of change goes far beyond policy, electoral, and legal battles. Major public marches and rallies continue across the islands, bringing together coalitions around general themes of "aloha 'āina," "reclaiming democracy," and "people over profit." Within communities that are most impacted, citizen science efforts are collecting more firm evidence about the extent of resident exposure to pesticides. Others have started to link to global activist networks and build international solidarity with places like Puerto Rico. A small group of activists traveled to Switzerland and interrupted a Syngenta shareholder meeting to denounce the company's poisoning of school children in Hawai'i. There is internal contestation within institutions like the University of Hawai'i's College of Tropical Agriculture and Human Resources. Progressive farmer and farmworker organizations that challenge the hegemony of the Farm Bureau are gaining a stronger presence. Many activists are most focused on developing alternatives to agrochemical-GMO agriculture in the islands, including a range of initiatives to expand agroecological farming. Some local agriculture projects are centered on Native Hawaiian rights to land, water, culture, and the resources and sovereignty to grow food. Decolonial and anticapitalist aims are implicit or explicit to varying degrees in such projects. These efforts—and many, many others—are part of a web of decentralized organizing that aims both to hold the agrochemical industry accountable for its local impacts and to challenge and change some of the root conditions that have created the situation in the first place.

In confronting root systemic inequalities and injustices, activists are part of a wider tide of leftist political organizing. Increasingly, solidarity is being built across intersecting struggles. Labor, economic, environmental, decolonization, demilitarization, feminist, and antiracist causes are finding common cause and creating spaces for organizing across and beyond issue silos. Events such as the People's Congress in 2016 and 2019 brought together hundreds of organizers from around the islands to "illuminate the root causes of injustice we all face." Such gatherings have led to coalitional work at the level of electoral and legislative politics—and beyond—that has not been witnessed in recent decades. While the need for systemic, intersectional struggle is much greater than what has been achieved, movements are clearly pushing in that direction.

As with all struggle, social justice movements in the islands face strong counterresistance from entrenched power. The agrochemical-seed-biotech industry benefits immensely not only from local-global political arrangements of imperial capitalism but also from its ideological pillars that discipline collective imagination of what is and is not possible in the social order. Mobilizing especially around neoliberal anti- and post-political ideology, the industry's countermovement in Hawai'i has significantly intensified since 2013.

Fig. 7. Thousands take to the streets of downtown Līhuʻe in the Mana March (September 8, 2013) to support Kauaʻi's Bill 2491, carrying signs with sayings such as "Pass da Bill, Not da Buck," "Power to the People," "Kauaʻi Deserves the Right to Know," and "Ua Mau ke Ea o ka ʻĀina i ka Pono" (the sovereignty of the land is perpetuated in righteousness, a phrase spoken by Kamehameha III upon restoration of the Kingdom of Hawaiʻi's independence from colonial British rule in 1843). (Photo by and reproduced with permission of Joel Guy, 2013)

Fig. 8. A group gathers at the start of the Kauaʻi Mana March behind the sign "Moms for Aloha ʻĀina," as well as several Kānaka Maoli flags. (Photo by and reproduced with permission of Dylan Hooser, 2013)

Fig. 9. Demonstrators at the Kauaʻi Mana March end at the Historic County Building, meeting place of the council, to press their demand for the passage of Bill 2491. (Photo by and reproduced with permission of Dylan Hooser, 2013)

Fig. 10. Following the agrochemical industry's lawsuit against three island counties over the passage of environmental regulations, protestors in hazmat suits drop banners at the Hawaii State Capitol on the opening day of the legislative session. Banners read: "Uphold the Public Trust Doctrine," "Syngenta Dow Pioneer: Shame on You Suing Kauaʻi County for the Right to Spray Poisons Next to Our School, Homes & Hospitals," "EVICT Monsanto Pioneer DuPont Syngenta Dow/Mycogen BASF," and "Monsanto Dow/Mycogen: Shame on You Suing Maui County to Throw out OUR Vote & Avoid Impact Studies" (Photo by and reproduced with permission of Elif Beall, 2015)

Fig. 11. Activists at opening day of Hawai'i state legislative session hold banners shaming the chemical industry for its lawsuits against three island counties. Kaua'i County Council member Gary Hooser, who co-introduced Bill 2491, speaks to the crowd. (Photo by and reproduced with permission of Elif Beall, 2015)

Fig. 12. Opening day of the Hawai'i state legislative session, which has become a space for intersectional social justice movements across the islands to gather. Signs read "Mana," "Aloha Aina," and "No GMO." (Photo by and reproduced with permission of Elif Beall, 2015)

Chapter 8

Battling Monsanto in an Era of Neoliberal Cynicism

As resistance to the agrochemical-seed-biotech industry has grown in Hawai'i, extensive resources and strategic energy have been put toward marginalizing the movement and reinforcing industry power. The industry and its allies have fiercely fought even the most conservative of demands for greater protection from pesticide use. They deploy wide-ranging methods to convince the public that change is unnecessary, undesirable, or unattainable. Debate is strongly influenced by key narratives of inevitability and allegations that the controversy is based on misinformed hysteria and self-serving political agendas. These narratives unfold within a wider landscape of neoliberal dogma, anti-political sentiment, and atomized cynicism. This ideological landscape has conditioned collective imagination of what is possible in the social order for decades. The corporate countermovement in Hawai'i provides valuable insight into how possibility is narrowed and disciplined from within particular sites, and why the battle over collective imagination is a most critical terrain of struggle for dreamers of a better world.

Disseminating Ideas to Block Change

Bill 2491 and subsequent county laws provoked major investment and coordination by the industry to crush the movement and the regulatory tide. Industry tactics include lobbying, campaign contributions, lawsuits, charity, infiltration of everything from business organizations to hospital boards, establishing strong influence in public schools and universities, and other quieter efforts within impacted communities to stamp out the voices of the discontented. Intertwined with these techniques, the industry and its allies employ a strategic network to produce and disseminate ideas in their favor. Multiple publications and organizations have sprung up since 2013, though many have disappeared as quickly as they appeared. With names like Farmers

& Friends, Family and Friends of Agriculture, and Kauai Farming and Jobs Coalition, none is forthcoming about being an industry appendage but all play an integral role in dispersing its narratives. Some have their own publications, while others pay "journalists" to write articles for newspapers, magazines, and weeklies. Publications and articles are often more subtle than direct industry PR or blogs but distinctly aligned in their narratives. Other organizations function more like astroturf groups, appearing to be broad and grassroots but actually conceived and funded by industry. Some of the new publications and astroturf groups seem to channel their funding through the Farm Bureau, designed for and with resources from the agrochemical industry. For instance, following the passage of county laws in 2014, an internal Hawai'i Farm Bureau memo announced a $400,000 annual budget for "a comprehensive public relations campaign" to counter "extreme activists" with "targeted messaging delivered by print, electronic (radio, TV and internet) and social media that utilizes guerrilla marketing techniques" (Manfredi 2014, 2).

Entirely overlapping with these efforts, several journalists, bloggers, academics, and industry employees from Hawai'i have joined the ranks of the Alliance for Science, located at Cornell University, as staff members, fellows, or associates. The Alliance for Science was launched and continues to be sustained by millions of dollars in Bill and Melinda Gates Foundation monies. The Alliance for Science functions not in isolation, but as part of a wider network of powerful actors aiming to shape the future of food and agriculture in increasingly corporatized and industrialized directions. The Gates Foundation has emerged as central to this agenda, spending over $375 million on agricultural development grants in 2019 alone (AGRA Watch 2020). They have especially sought to expand corporate agribusiness in Africa, forcefully pushing a "Green Revolution for Africa" similar in form to the twentieth-century Green Revolution that fostered the Global South's dependency on commercial seeds, chemicals, fertilizers, and machinery from the United States (Patel 2013). While the Gates Foundation's agricultural investments and corporate agribusiness partnerships have come under increased scrutiny (Gates Foundation deputy director Rob Horsch himself worked for Monsanto for twenty-five years), their framing of the debate about biotechnology and capitalist agribusiness more generally has been more disguised and arguably deceptive (AGRA Watch 2020).

The Gates-funded Alliance for Science was founded with the purported goal to "depolarize the charged debate" about GMOs, and it describes its primary strategies as communications, training, and establishing a global network. While claiming neutrality, its messages are strikingly aligned with industry talking points on GMOs and pesticides. Emails obtained through a

Freedom of Information Act request and posted in the UCSF Chemical Industry Documents archive show close coordination between the agrochemical-biotech industry and CAS on key public relations initiatives. So close are the relationships and communication efforts that the Alliance for Science is accused of being "an aggressive propaganda tool for corporate biotech and agribusiness" by journalist Stacy Malkan (Malkan 2016; AGRA Watch 2020). The Community Alliance for Social Justice writes that the Alliance for Science is "effectively doing the work of corporations under the veil of philanthropic benevolence" (AGRA Watch 2020, 6). Through their affiliation with a prestigious university, the Alliance for Science masquerades industry talking points as "scientific" while in fact distorting scientific debate about biotechnology and agrochemicals. Beyond promoting uncritical, blanket acceptance of biotechnology, the Alliance for Science goes further to marginalize alternatives, including likening agroecology to being "anti-science" (AGRA Watch 2020, 13). Given the increasingly widespread recognition of the need for agroecological methods to deal with growing environmental crises—reflected, for instance, in the International Assessment of Agricultural Knowledge, Science and Technology for Development (see chapter 2)—this is a blatant misrepresentation. It is no coincidence that it comes at a time when there is "mounting global scientific consensus around the merits of agroecology," a fact that challenges the industrialized capitalist model of agriculture at its very core (AGRA Watch 2020, 19).

One of the primary programs of the Alliance for Science is training and supporting global research fellows to, in the words of Stacy Malkan, "conduct public relations and political advocacy that aligns with the agrichemical industry's agenda" (Malkan 2016; see also AGRA Watch 2020). Fellows have backgrounds in journalism, marketing, business, academia, and government and are from places where the agrochemical-biotech industry' presence is most controversial, including Hawai'i. Consistent with the Gates Foundation's Green Revolution for Africa goals, over 60 percent of the Alliance for Science's global research fellows in 2019 were from Africa (AGRA Watch 2020). Individuals from Hawai'i who are associated with the Alliance for Science blog, publish, and comment prolifically on media articles. It appears that for at least a few individuals, this has become a full-time, paid job. Daily blogs and editorializing most frequently consist of highly personalized attacks on activists, health care professionals, scientists, and policymakers who raise questions about industry pesticide use. Some of these same individuals are also "independent experts" for GMO Answers, a marketing website created by Ketchum, a PR firm, for the industry. Alliance for Science's global research fellows and GMO Answers experts gain legitimacy in the eyes of the public by ostensibly being separate from the industry and Gates Foundation. As the Community Alliance for Social

Justice points out, this adds power to the narratives coming out of the Alliance for Science and is a "pernicious method" for influencing "communications, narratives, and policies" around the future of food and agriculture (AGRA Watch 2020, 6).

In Hawai'i, through mechanisms like the Alliance for Science and others, a small group of actors has created a communications infrastructure that appears diverse and delinked, but is actually highly deliberate and coordinated. Though there are notable differences between the actors that contribute to this communications network, all are indisputably tied to industry initiatives. The resources behind these efforts are vast. Just on reported direct advertising against Maui County's ballot initiative, Monsanto and Dow spent around $8 million, more than any group or individual has spent in any Hawai'i election (Kerr 2014). This included dozens of thirty- and sixty-second television and radio spots, 30-minute television infomercials, newspaper ads, mailers, and online advertising on sites like Pandora. The reported spending against Maui's ballot initiative does not include all of the subtler and ongoing communications tactics described above. The money spent in Hawai'i is only a fraction of what is being spent globally on ongoing efforts like GMO Answers and the many industry front groups that are constantly at work waging a countermovement against health and environmental advocates and scientists.

Identifying industry mechanisms of propaganda does not mean that industry and its close allies are the sole source of ideas that serve its interests. This is far from the case. In what follows, common industry-bolstering narratives are summarized and analyzed. However, these emerge from a wider neoliberal moment. They are often reproduced in subtler ways by those who claim skepticism of the industry and allegiance to social justice but see only imperfection and contradiction in collective struggle for social change. Thus, the intent here is not to categorize or dismiss the narratives mapped below as merely industry myths. Instead, it is to show how they ultimately function to foreclose change and hold in place injustice. They are part of a much broader ideological landscape and provide valuable insight into the depth of neoliberal depoliticizing logic as it functions to shape public common sense. There are lessons here, then, not just for Hawai'i's struggle but for every struggle facing the pernicious doctrine that real change is impossible.

Key Narratives against Change

Quotes and verbiage below are from blogs, opinion pieces in newspapers, social media, comments sections in online publications, public testimony (written and oral), and statements by politicians. Unspecified quotations are common remarks in media or other visible spaces of public conversation. These are among the most developed and extensive industry-bolstering narratives consistently appearing across multiple forums. They are drawn most especially from the original controversy over Kauaʻi's Bill 2491 in 2013. Analysis follows a short summary of three prominent, overlapping narratives.

"We Just Can't"

Both at the local and global level, the industry presents itself as that which we cannot live without. Globally, feeding the world and modern society are said to be made possible through industrialized capitalist agribusiness and the agrochemical-seed-biotech industry's products in particular. Locally, the industry professes to provide irreplaceable jobs, taxes, economic ripple effects, investment in infrastructure, and support to small local farmers whose supply chains would collapse if not for their presence (Loudat and Kasturi 2013; Misalucha 2015). It is insisted that any regulation of the industry will provoke a chain of negative consequences: "Bill 2491 will destroy local agriculture," "it will hurt farmers in poor countries" (Save Kauaʻi Farms 2013), and it will "cripple" the economy. Testimony against Bill 2491 speculated that the pesticide disclosure law could lead to the closure of hospitals and schools, luxury hotels replacing agricultural lands, the county going bankrupt, and the loss of individual freedom in the slippery slope of government intrusion. Maui's ballot initiative provoked an even higher pitch of catastrophism. It was repetitiously stated that the "Farming Ban" would completely devastate the county's economy and cause widespread unemployment.

"Politics Are Disrupting Aloha"

Amid grand proclamations about the disasters of pesticide regulation, those who raise health and environmental concerns are said to be fomenting divisiveness. Activists are accused of "deliberate fear-mongering" and "manipulation . . . to advance a political agenda" (Conrow). A "radical minority taking over" is blamed for "dividing the community" and destroying the "spirit of aloha" (Take Back Kauai Facebook group; Aole GMO Means Aole Aloha n.d.). Violence and juvenility are associated with the "No GMO = No Aloha movement," including frequent reference to "mobs" and behaviors described as "rude," "threatening," "temper tantrums," "rabid," "vandals," "spoiled brats," "yelling and screaming," "obnoxious," "immature," and "low-level discourse" (Eagle 2014; Kamiya; Conrow). In particular, "outsiders" or "newbies" with a "cynical agenda" are accused of driving the disruption (Kester 2014; Conrow). In the "war" against the chemical industry by those engaged in "group think totalitarianism" it is said that not only is the industry victim, but so are its workers, the wider community, and local culture (Conrow). Save Kaua'i Farms, an industry astroturf group, proclaims "We fight for our friends, we fight for our jobs, we fight for our families, *we fight for our way of life*" (2013).

"Overcoming Politics and Hysteria"

Those purportedly driving the division by raising health and environmental concerns are disparaged as driven by emotion, fear, and "politics," as opposed to science, rationality, and fact. Activists are branded as hysterical, fanatical, Luddite, paranoid, delusional, and conspiracy theorists (Conrow; Harmon 2014; Kamiya; Kloor 2012). In contrast, the industry claims to offer a "science-based perspective." As activists are repudiated for being perpetrators of divisiveness and makers of myth, the industry claims desire for collaboration and the expert knowledge to "move forward together." It is widely asserted that "common middle ground" can be found through "cooperation" and "nonpolitical" solutions. Such solutions are said to lie in voluntary action, "education" of the public, and "neutral fact finding" that does not involve "politics" (County of Kaua'i Resolution 2013-72). The restoration of aloha, goodwill, and local ways and values will be realized by eradicating politics and rabble-rousers: "Many of the anti GMO folks are using the bill as a cause of strife to drive out the corn companies. Their anger and hysteria are causing a great erosion of our aloha spirit. We have a duty to protect and nurture our aloha spirit, so we hope that you will help to restore the Hawaiian value of *Lokahi:* unity and harmony, by working cooperatively toward agreement instead of litigation" (Public testimony, Bill 2491).

Together these intertwined narratives of inevitability, objectionable politics, ignorant masses, and the favorability of nonregulatory "solutions" work to subvert resistance and meaningful change. The rest of the chapter turns to examine these within the wider ideological landscape.

The Inevitability of Agrochemical-Seed-Biotech Corporations

What unfolds in the organization of human societies is in part conditioned upon collective belief of what is possible. In the words of anthropologist David Graeber, humans "make and maintain reality" through shared imaginations of capabilities and potentialities (Graeber 2009, 523). The history of human societies, of social relations and structures, is one of monumental, dynamic, and untold change. Despite this historical truth, for decades neoliberalism has told the story of "this is as good as it gets." Fundamental alternatives to a social order in which astonishing numbers of people live in utter deprivation and die from lack of access to available resources is largely treated as fantasy. In general, such a system—capitalism—does not gain legitimacy by presenting as ideal. Instead, at local and global scales, and through a variety of intertwined ideological and material mechanisms, the mantra of the past decades that "there is no alternative" is continually reproduced. A historically specific and relatively recent social system structured by competitive accumulation of private wealth is naturalized as all there is and all there could be, save perhaps totalitarianism. While struggles against inequality and environmental ruin abound, there is at the same time a dominant acceptance of a social order that produces destitution and is leading to planetary apocalypse.

Perhaps above all else, the neoliberal project of the past decades worked to devastate the collective imagination (Graeber 2011, 6). Though starting as a fringe intellectual movement, neoliberalism grew rapidly into a starkly dominant imaginary of the world (Peck and Tickell 2002, 381). Mark Fisher describes the collective psychic atmosphere taking hold especially since the 1980s as "capitalist realism" (2009). There was a widespread sense that it is impossible to imagine a coherent alternative not only to capitalism, but to global monopoly-finance capitalism dictated by a smaller and smaller corporate elite. While Margaret Thatcher's proclamation of no alternative to neoliberal capitalism was originally stated as a matter of viability, by the mid-1990s it came to mean no imaginable alternative to the marketization of almost all aspects of life and attendant domination by monopoly corporations (Fisher and Gilbert 2013; Brown 2015). It came to be simply common sense that everything in society—from seed reproduction to healthcare to scientific advancement to education—should be run as a business, dictated by competitive profiteering by the already most powerful capitalists (Fisher 2009, 17).

At the time of this book's writing, major events are happening that signal the possibility of significant breaks in neoliberalism's colonization of the imagination. Public policy debates are shifting to a remarkable degree, and the neoliberal "center" has largely lost its authority. As leftist challenges to neoliberalism ramp up globally, deeply entrenched power and wealth is set on branding any challenge to the status quo as impossible, dangerous, or some other instantiation of unthinkable. Across a wide variety of contexts, whether in bailing out banks, slashing government services, privatizing public goods, deregulating corporate harms, or enabling corporate consolidation, a similar hyperbolic logic runs through: we cannot do it any differently; there are no other options; the market will not permit anything else; things will fall apart if we try anything new. These mythologies are observable in Hawai'i's agrochemical industry conflict, where they take shape particular to context but are undergirded by consistent neoliberal rationale of inevitability. While 2021 is already a very different ideological terrain than 2013 was, there are continuities and lessons that reach back far beyond the advent of neoliberalism.

When in 1869 Hawai'i sugar baron Samuel Castle responded to critics of the contract labor system as "striking a serious blow at every interest in the country" (1869, 3), he gave clear illustration of mechanisms by which capitalist injustice is constructed as inevitable. Indeed, by certain indicators it was true, as a 1905 congressional study remarked, that "directly or indirectly, all individuals in the Territory of Hawai'i are ultimately dependent upon the sugar industry" (in Baker 1911, 28). The fundamental question too often unasked—and always easier to ask of the past than of the present—are the means by which dependence on not only the plantation economy but to the plantation oligarchy was manufactured, sustained, and tangled up in ways that gave the appearance of indistinguishability between common goods and select private interests.

In today's Hawai'i, most residents—from the lowest-paid fast-food workers to oligarch-descendant landowners collecting hefty rents—are reliant on the steady flow of millions of visitors annually. Even in dependence, it is well understood that Hawai'i's model of corporate tourism creates incredible vulnerabilities; indeed, this has been the harsh lived reality of Hawai'i's working class through the COVID-19 pandemic. With daily arrivals of travelers dropping from around thirty thousand per day to five hundred per day, joblessness rose to Great Depression levels in 2020 (Finnerty 2020). Even in normal times, corporate tourism is an economy that offers low-paying jobs but drives up the cost of living, resulting in high rates of poverty, homelessness, and displacement from the islands. The compulsions of endless development and rural gentrification and the huge sacrifices made to the environment and valued "island lifestyle" are widely lamented facts even among those who benefit financially. In a severely narrowed discourse of social possibility, the solutions to such systemic problems are most typically presented as better planning and creation of more elite tourism markets. The search for "economic diversification" remains dominantly defined as facilitating new forms of overseas corporate investment (CEDS 2010). From within this extremely limited horizon of possibility, agrochemical operations appear a welcome diversification, a contribution to the economy that cannot be sacrificed, and the most important preserver of agriculture in its loss to development.

In celebration of agrochemical corporations as saviors of agriculture and livelihoods, a dominant narrative of their necessity and inevitability is established. This narrative assumes as immutable the social arrangements that give shape to Hawai'i's monoeconomy and make transnational corporate agribusiness the most viable addition. Beyond larger forces of capitalism and imperialism, such arrangements include numerous forms of local public subsidy that exacerbate and extend capitalism's general privatization of benefits and socialization of costs (see chapter 6). Local and global conditions that could be different are obfuscated. The fact that corporate agribusiness systematically displaces alternatives is taken as just the natural order of things. Tales of a plantation past that justify the need for "Big Ag" selectively ignore the direct marginalization and often violent displacement of alternatives, including the near entire devotion of natural resources and government treasury to particular business interests. As this history repeats in corporate tourism and chemical companies, desires for something different are treated as naive misunderstanding of how the world works.

In the tale of inevitability and impossibility, limitations from within the conditions of the present are taken to represent the entirety of social possibility. An endless list of constraints—state tax dependencies, a shortage of public funds, a population lacking the skills for different economies—come to signify the whole of the situation and its unchangeable nature. In what Henry Giroux describes as a "discourse of disconnection," issues are fragmented and treated as isolated dilemmas (2014, 65). This narrows collective capacity to perceive and respond to structurally and ideologically inseparable socio-economic-cultural problems. When treated as an isolated dilemma, dominant ideas of agrochemical industry inevitability are accurate. It is true that only very particular forms of capitalist agribusiness can outcompete other capitalist agribusiness within particular market and political arrangements. When detached from other systemic arrangements that must also be changed, it appears true that only agrochemical companies can sustain local livelihoods, infrastructures, and agricultural lands.

In one common example, agrochemical operations are described as necessary to sustaining small local agriculture, the "anchor" of all agricultural potential in the islands (Misalucha 2015). Evidence for such a claim is taken from the fact that they lease large amounts of agricultural lands and thus keep them from being developed, that they sublease land to local farmers, that they maintain irrigation and other infrastructure, that they purchase agricultural products at a scale to stimulate local markets for other farmers, and that they fund agricultural scholarships and institutions (Hervey 2012; Gordines 2013). To take just one of these, access to affordable long-term land leases is one of the most substantial barriers to local food production in the islands. Causes for lack of accessibility include interrelated factors of highly inflated land costs, speculation and development, consolidated land control, land use and zoning laws, and state land leases that seem to automatically go to transnational agribusiness—in short, poorly regulated capitalist land markets in a development-based economy and US dispossession of Hawaiian lands. Without considering root causes and how they could be addressed, land subleases to local farmers from agrochemical corporations appear a welcome gift. They seem the best—or only—option when others are off the table. Similarly, scholarships from Monsanto are a received enthusiastically when it is assumed that public institutions will continue to be impoverished. The intention in these examples is not to dispute the confines or the complexity of working within the limitations of the present. Rather, the point is that when broader and deeper change is thought to be impossible, the situation can appear to be highly constricted and immutable, and agrochemical offerings to be necessary or a best alternative.

At a more surface level, ideas of "no alternative" dismiss even immediately available, relatively straightforward possibilities for a more ecologically regenerative local agricultural economy that supplies decent livelihoods. Examples of immediate possibilities include making stolen public lands available to Hawaiians for their own food sovereignty, making public resource and subsidy supports available to local producers rather than corporate agribusiness and tourism, shifting public university research funding to enhance agroecological food production and farmer education, creating stable local food markets by anchoring them to state institutions likes schools and prisons, incentivizing and supporting worker cooperatives through policy, enforcing existing public trust water law, changing and enforcing land use policy to encourage local agriculture, anti-speculative land taxes and more equitable taxes generally, raising the minimum wage, strengthening labor protection laws, and so much more. Such options, and many others, are frequently cited as too difficult, doomed to fail, potentially causing some unforeseen problem, or some other instantiation of "not possible." To this can be added the pitched fatalism around pesticide disclosure and buffer zone bills that allegedly would cause economic collapse, the demise of agriculture in the islands, and government intrusion into personal liberties.

Overlying localized claims of there being "no alternative" to agrochemical-GMO plantations are global ideas about monopolistic corporate agribusiness as imperative to feeding the world and making modern society possible. As with capitalist realism more generally, there is a sense that one does not have to like the consequences of a corporate capitalist food system, but society cannot live without it (Fisher 2009). At the core of such realism is the ideology that a capitalist market compels agricultural productivity and thus social progress more generally. Such dogma originated most especially in the enclosure movement in England, which saw peasants systematically dispossessed of common lands to make way for landlord profits. Lands were said to be "improved" from previous "waste" with the eviction of peasants and the commercialization of agriculture. This "improvement" was justified as being in the common good, as commercialization would purportedly lead to increased grain yields and thus modernization. Such rationale also vindicated global colonial endeavors and persisted in Hawai'i's plantation development and capitalist enclosures 150 years ago. Only slightly repackaged today, the restructuring of the food system toward increased privatization, corporatization, financialization, and industrialization takes place within a moralized discourse of increasing grain yields to "feed the world." Within this, private sector technological development is made synonymous with "social progress" (Newfield 2008). The agricultural biotechnology project is wholly immersed in such constructions of solving hunger through capitalist markets, innovations, and technologies.

As a place central to development of such technologies, Hawai'i is imagined as playing a virtuous role in feeding the world: "If Monsanto is pushed off this island . . . it will affect people here, and around the globe. This company discovers more productive ways to grow healthy crops from the rich, fertile soils in Iowa to the hard-baked lands of South Africa. Monsanto seeds survive droughts and hardships to feed multitudes. For me personally, I will lose a job I feel has allowed me to impact the world for good, and helped me to ease the suffering of many people" (Monsanto nursery supervisor, in Stoltzfus 2014). In Hawai'i, as elsewhere, "privileged" and "well-fed" activists are accused of being implicated in the starvation and death of poor people who are being denied technologies that would feed them (see also Chopra 2015). In another typical example, a widely circulated article by Alliance for Science fellow Nassib Mugwanya—"After Agroecology: Why Traditional Agricultural Practices Can't Transform African Agriculture"—blames Westerners for imposing an agenda that keeps Africans hungry and African farmers "bound to the soil and confined to poverty" by denying them biotechnology and "modernization" (Mugwanya 2019; see also AGRA Watch 2020, 15). The invocation of morality works to silence debate, while the ideological dominance of a particular capitalist trajectory of progress and productivity makes it difficult to challenge such an accusation (McAfee 2003, 215). The problem is not with productivity or technological innovation in their own right. It is what is justified, neglected, and silenced by such logic, including most especially the use of a moral discourse to code profoundly immoral projects. Many of the very projects claiming to increase agricultural productivity and innovation—from land grabs, to free trade, to seed patents—are directly implicated in the causes of global hunger and are more fundamentally about privatization, commodification, and accumulation of private profit for the already wealthy (De Schutter 2010, 2011b; Magdoff, Foster, and Buttel 2000; Lawrence, Lyons, and Wallington 2010; Holt-Gimenez and Patel 2012). Though premised on falsehoods, the global discourse of "necessary" corporate monopolies and technologies intertwines powerfully with local narratives of "no alternative" to buttress a sense of the inevitability of agrochemical plantations.

Nasty Activists and Dirty Politics

The severe contraction of what is considered possible in the social order is accompanied by suspicion and rejection of collective political engagement to change the social order. Disengagement from, and even hostility toward, the realm of politics intensified alongside neoliberalism's anti-statism and individualism. A culture of atomized cynicism has largely taken grip, one that disavows collective struggle for social change and consequently avows the impossibility of change. There is significant and growing motion in another direction today. Social movements of the past several years, especially those led by younger generations, have begun to break through the neoliberal rejection of collective political engagement. We are at a critical juncture, one that requires attention to how anti-political ideology has functioned in the recent past and continues to operate to subvert change.

Neoliberal ideologies of individualism largely collapse notions of social solidarity and collective struggle for change. When extreme individualism shapes understanding of the world, action for and by collectives is easily made unintelligible. Those who engage in collective struggle are regarded with suspicion and disapproval. Social movements are rebranded as mobs with divisive behavior and self-serving agendas. Rather than the historical force necessary for social change and justice, they are perceived as unruly, populated by objectionable characters, and prone to what Hawai'i's movement critics describe the "totalitarianism" of "groupthink" (Conrow). The ethos of individualism is not only a barrier to collective action; it disparages it.

The scorning of collective political action has been without pause in Hawai'i's controversy. Following a blog post by Gary Hooser celebrating people joining their "million little fists" to demand change, "fistee" was coined and branded as an indictment of "jackboot mentality." Social media and even newspaper articles erupted in disparaging remarks about the groupthink of fistees.

In this and innumerable other examples, the diverse people and ideas that converge around common goals for social change are treated as monoliths. Moreover, they are reduced to their most objectionable characters. In Hawai'i's movement, conspiracy theories, chemtrails, and reactionary fringes come to stand in for the whole. Every social movement has its inconsistencies and less-than-liberatory threads—these should not be ignored. However, characterizing entire movements by what is perceived as their imperfections or contradictions only serves to delegitimize their liberatory aspirations and is in large part an intentional strategy.

What is most repudiated is that which is tagged unruly and disruptive, or outside of the boundaries of proper participation in institutional politics (Khanna 2012). Young activists especially have been strongly criticized for "ugly," "disrespectful," "juvenile," and even "violent" behavior. Such fierce accusation has been issued for actions like speaking loudly and out of turn at council meetings. Conflictual social media discussions between small groups of people have also been selectively presented to depict a hostile movement. Selective and distorted representation is a classic strategy of vilification and marginalization, propagated by both corporate reactionism and media sensationalism. In Hawai'i, confrontational social media or other encounters are almost entirely highlighted by the industry and associated bloggers before seeping into wider public conversation. The verbiage of this popular blogger, who is employed by the Alliance for Science, has become widespread: "Their [activist] comments, though distasteful to thoughtful, intelligent readers, speak volumes about their immaturity, their unwillingness to accept facts, their dogmatism, their speculation and ignorance and their sheer nastiness. This is the core of the anti-AG/anti-GMO movement, in all its ugliness revealed" (Conrow). During and following deliberations over Bill 2491, the "ugliness" and "divisiveness" of activists was more discussed in media than the actual policy debate. The problem was increasingly recast as controversy (i.e., activism) itself. Rather than bringing to light conflict that already exists in the social order, activists are seen as the creators of conflict themselves. They are denounced for violence when seeking to rectify the violence of structural inequality and injustice.

Rejection of contemporary social movements frequently contrasts them to those of the past, portraying historical struggles as respectable and righteous but those of today as having taken an objectionable turn. In this depiction, history is rewritten. The militancy, radicalism, and disruptiveness of past movements is erased, including transformative labor and anticolonial and antiracist struggles in Hawai'i. Resistance is said to belong in the past, to the days of sugar oligarchs and indentured servitude. There is widespread celebration for having supposedly achieved a more equitable, multicultural "era of consensus" following the Democratic Revolution (Kent 1993, 122). An article in *Farmers & Friends*—a publication initiated by agrochemical-seed-biotech industry collaborators—narrates this mythologized harmony and its troublesome disruption by activists:

> Not so long ago, progressives in Hawai'i believed in progress. They thought that innovation boosted the human prospect. Call it the "Burns Years," post-statehood. There was a lively sense that Hawai'i's people could accomplish great things as they shed the plantation era yoke of "subtle inferiority." As equality took root, an easygoing collaborative style blossomed. It tasked the Democratic Party to develop a new, diversified economy . . . People might disagree—and often did—but it was understood that bad blood was bad manners. People knew when to say, "Eh, no act, brah."
>
> Now, a new variety of progressive politics has seized Hawai'i. It is confrontational and employs the organizing logic of California-style initiatives. Sign up like-minded true believers on Facebook and storm city hall with legions of testifiers. Its energy is self-replicating and it is enjoying a heyday . . . Rather than sort through our differences via dialogue, real fact-finding and problem-solving, we are playing guerrilla theater war games. (Flynn 2014)

Denouncement of the purportedly "confrontational," "divisive," "guerrilla war" style of contemporary movements is especially tied to neoliberal ideas about individualism. Hyper-individualistic ideology assumes that people are innately self-interested. Such ideology renders solidarity and struggle for and with others unfathomable. Rather than altruistic concern, Hawai'i's environmental and health justice activists are accused of being driven by personal gain, a desire for conflict, or some hidden agenda. Again, such allegations are a widespread and well-practiced corporate tactic used to discredit activists. In an ultimate distraction, collective political goals are made personal.

Self-serving and divisive activists are accused of "attacking" not only agrochemical-seed-biotech companies but also workers, their families, farmers, agriculture, and a local "way of life" (Save Kaua'i Farms 2013). Moreover, attacks on corporations are attacks on the poor and hungry, modern society and progress, science and intellect (Manfredi 2014; see also Dibden, Gibbs, and Cocklin 2013). A Hawai'i Farm Bureau warning to its members about proposed pesticide disclosure regulations that would have applied only to the largest users reads:

> Across the nation, farmers and ranchers have been caught off guard by the extremist activists that will stop at nothing to realize their utopian, misinformed and unsustainable visions of how you should farm . . . [F]armers and ranchers are demonized to the point where young people are further dissuaded from entering careers in agriculture—while activists selfishly attempt to advance their own agendas and political careers . . . [T]hese efforts are aimed at driving modern production agriculture out of business . . . [I]f successful [they] would destroy more than two centuries of evolution in agriculture practices in Hawaiʻi . . . Its success would cause the collapse of agriculture in Hawaiʻi and have reaching impacts beyond our shores. (Manfredi 2014, 2–3)

Organizations in Hawaiʻi, some part of larger national organizations, similarly face widespread allegations of "hidden agendas of wealthy outside philanthropists," "intruding in local politics," disguising an "industry versus industry battle," being financed by real estate agents with cynical motivations, "hijacking" the environmental movement, and "well-funded campaigns of misinformation" (Hofschneider and Grube 2014; DePledge 2014; Manfredi 2014; Conrow). Blogs and news stories have given extensive coverage to "following the money" in the "anti-GMO" movement (Hofschneider and Grube 2014; DePledge 2014; Entine 2013; Kamiya; Conrow; Farmers4Choice Facebook group). Such accusations reduce a diverse, broad, and decentralized social movement to the activities of a handful of organizations. Moreover, the magnified attention on a few organizations' funding sources obscures extreme differences in resources. Contradictorily—and revealing of general consent toward corporate power—it is largely assumed that industry will spend tens of millions of dollars on political campaigns, propaganda, and lobbyists, while those who challenge them must be resourceless to maintain their morality.

It is true that there are wide-ranging and deleterious impacts of funding structures and the nonprofit industrial complex on social justice movements. Activists face real structural challenges and contradictions in trying to establish sustaining modes of organization that can carry them forward. However, critique of the nonprofit industrial complex is also deliberately manipulated to nurture cynicism about broad social movements. In the words of corporate strategist Richard Berman (aka "Dr. Evil"), one of the most potent tools of the elite is to "diminish the moral authority" of activists and foster public distrust of "political agendas" by targeting "do-gooder, left-wing organizations" (Berman 2014). The extensiveness of these corporate campaigns has been revealed in struggles ranging from Standing Rock and climate justice, to reproductive rights, to union fights.

In a wider ideological landscape that disparages activism and collective struggle, industry smear campaigns find enthusiastic audiences and have widespread effects in marginalizing, silencing, and depoliticizing. Personal slander tactics create a chilling effect that is hideously antidemocratic (Hager 2014). They work to defame and distort, and they contribute to more general disaffection. They need to be understood as part of a wider phenomenon undergirded by anti-political and individualistic rationalities, one that powerful interests have found extremely effective for stifling dissent and change (Oreskes and Conway 2010).

What is presented as preferable to collective engagement, politics, and dissent is isolated, apolitical, individualized action. Margaret Thatcher's proclamation that "there is no alternative" to hypercapitalism went hand in hand with her declaration that "there is no such thing as society," only individuals. Neoliberal ideology seeks to replace ideas of collectivity with a belief that our only duty is to "look after ourselves," in the words of the late prime minister. Social change is said to happen not through collective engagement to transform systems and policies, but by "looking within" to alter individual consciousness and lifestyle. The same Alliance for Science–paid blogger who frequently chastises the "ignorance," "nastiness," and "ugliness" of Hawai'i's "anti-GMO movement" instructs: "We can't blame government, or even corporations, because they are merely a reflection of us. We've created them through our belief systems, our voting, our consuming and lifestyle choices. Nor can we expect those entities to fix anything for us. We've got to engage in the hard, dirty work of consciousness change ourselves" (Conrow). By such logic, there is no social structure, no systemic injustice, no power imbalance in the existing order that must be challenged. Rather, there are only free-floating individuals who are all complicit in societies' ills merely by virtue of being born into the world as it is. The assumption is also, then, that if one cannot fully escape global chains of capitalist exploitation, then one is not entitled to go about trying to change them. Thus, the only rational action is to address one's personal lifestyle choices. To attempt wider sociopolitical change is chided as an inescapably self-righteous and duplicitous act. In opposition to the intrinsic imperfections and contradictions of collective social struggle, the ethical individual focused on their personal lifestyle is acclaimed. Movements—messy, diverse assemblages converging around common social goals—are expunged from history as the only force that has ever truly remedied injustice (Zinn 2012, 293).

From Divisive Politics to Consensus and Unity

The disavowal of collective engagement, of people's movements to change an unjust society, is accompanied by a related post-political ideological trend. Real democratic politics is the ability to debate, to criticize the existing order and challenge it. Real politics recognizes that there are competing interests at stake and thus necessitates accepting conflict (Swyngedouw 2007, 25). In contrast, post-political ideology begins from the mythology of consensus and cooperation. It suggests that there is no need for real politics, that the world can be mediated by the simple coordination of facts and interests. As such, it depoliticizes what are critical political matters. Conflict with and actual change in the existing order is pacified when politics is staged as merely the management of consensus. While some discussion and dispute are encouraged, it is only permitted within a narrow frame in which problems are ultimately resolved through techno-managerial production of consensus (Swyngedouw 2011, 77). Approaches aiming to operate outside of politics or build apolitical "win-win solutions" to problems are applauded, while those that confront power and seek to actually change the social order are vilified as divisive.

Divisive, confrontational, and antagonistic politics are not a thing left behind in what some call a post-political turn to consensual management. Illustrated in Hawai'i's example, it might even be argued that they are most characteristic of a negative, personalized, and largely informational politics that descends into sound bites and scandals (Castells 2005, 162). Rather than substantive debate and controversy about policy and system design, tweets and personality dramas take center stage. Hateful commentary blanketing the internet and the proliferation of smear tactics are illustrative of the personalized attack politics that have become so pervasive (Rohrer 2012; Mulvey and Shulman 2015). Sheldon Wolin argued that it is the very trivialization of political discourse that stirs fragmented emotional outrage while leaving power and the social order unchallenged. For Wolin, "It is all politics all of the time but a politics largely untempered by the political" (2008, 66). There is continuous squabbling between political parties, corporate powers, and rival media outlets, but what is absent is the political—the collective engagement in "finding where the common good lies amidst the welter of well-financed, highly organized, single-minded interests" (Wolin 2008, 66).

Similarly, in Rancière's thinking on post-political ideology, "the essence of consensus is not peaceful discussion and reasonable agreement as opposed to conflict or violence. Its essence is the annulment of dissensus" (2010, 42). It is the containment of all disagreement to fit within the existing social order, within "the partition of the sensible" (Rancière 2010, 36). Rancière writes: "A culture of consensus is accordingly created that repudiates the old conflicts, accustoms us to dispassionately objectivizing both the short- and long-term problems that societies encounter, to asking experts for solutions, and to discussing them with representatives qualified in grand social interests" (2006, 75).

The dominance of post-political ideology has certainly been challenged in the past years' events of collapsing neoliberal "centrism" (an ideological term for a political agenda that has by no means been moderate, historically speaking). Polarized political discourse is the norm, though it remains more in the realm of what Wolin describes as trivialization that stirs outrage while leaving untouched the big political questions about how to organize our lives and whose interests will be most served. Still, the continuing appeal of post-political ideas, and their ability to depoliticize (and thus foreclose) what are necessarily political conflicts, should not be underestimated. These theoretical observations remain highly relevant to Hawai'i's agrochemical controversy, especially as it exploded in 2013 and 2014.

With the initial outburst of politics around Bill 2491, seductive post-political ideas worked to distract from actual change. Industry's "solutions"—largely supported by the state—advertised appealing prospects of "working together," "middle-ground," "nonpolitical and science-based perspectives," and "better education of the public." All were tied to voluntarism. Just days before a critical vote on Kaua'i's pesticide disclosure law, Governor Neil Abercrombie announced that the state would "be putting forth standards and guidelines for seed/diversified agriculture companies to voluntarily comply with certain health and safety requests of the community" (Abercrombie 2013). An undefined "Kaua'i delegation" would lead the process. None of the impacted residents, activists, or policymakers involved in advocating for Bill 2491 were included as part of the delegation or ever consulted for feedback on the proposed "steps to address the concerns of the community" (Abercrombie 2013). In an entirely nontransparent and antidemocratic process, its happenings were made public only after the industry and state had met to craft their version of a solution in a voluntary pesticide reporting scheme, the Kaua'i Agricultural Good Neighbor Program. The voluntary program was widely criticized by concerned community members as an attempt to derail county regulatory action. However, the state proceeded, introducing its proposed program on the eve of a critical county council vote over Bill 2491.

The Kaua'i Agricultural Good Neighbor Program was voluntary, included only highly generalized monthly reporting of restricted-use pesticides, specifically exempted reporting that might compromise companies' trade secrets, and requested buffer zones of one hundred feet around schools and hospitals. Its preapplication notices of spraying went to only a handful of schools and hospitals and were so generalized as to be useless. In contrast, Bill 2491 mandated legally enforceable weekly disclosure of all pesticide use, including precisely when, where, and what was sprayed and the wind speed and direction at the time of spraying. The county law additionally required 500-foot buffer zones around homes, schools, and medical facilities, 100-foot buffer zones around waterways, detailed preapplication notification of spraying for all residents within 1,500 feet of an operation's property line, a provision for health care providers to access information about acute exposures, and an independently conducted health and environmental impact study. Despite clear differences, the voluntary Good Neighbor program was consistently referenced as evidence of why democratically deliberated regulatory action is unnecessary and what type of consensus should substitute for politics. As agrochemical companies sued to block implementation of laws, their participation in the voluntary pesticide disclosure program was hailed as evidence of their "spirit of collaboration" (Nagaoka 2014). Syngenta stated, "the fact that we are doing this on a voluntary basis underscores our commitment to transparency and keeping the community informed" (Phillipson 2015, 1).

In 2015, following major struggles at the State Capitol over pesticide bills (none of which passed), there was again a push to extend voluntary pesticide reporting statewide. The proposed program was designed by industry and the Hawai'i Department of Agriculture. Congruent with industry, the Department of Agriculture has lobbied extensively against regulations at the same time as they move to implement a voluntary program. Instead of democratic deliberation, industry and the state vowed to share what they had decided "once we have something in place." Their proposal focused especially on expanding "our education and outreach efforts to ensure the public that we are using the most updated farm stewardship practices to care for the health of our neighbors, communities and our land" (Hofschneider 2015c). Rather than change industry practices, the emphasis is on changing how the public understands them. Always held in tandem in this industry-state discourse is disregard for concerns and prescription of the untransparent and undemocratic "consensus" solution to such concerns. Making the public more amendable to industry "facts" is typically part of these solutions. In essence, the position posits that the masses are largely ignorant, and thus democracy should be terminated in favor of corporate-state administered education and decisions.

The Kaua'i Agricultural Good Neighbor Program quietly disappeared after twenty months. The statewide good neighbor proposal got a lot of chatter when regulatory options were on the table but never materialized. Their outcomes are clear testament to the goals of voluntarism. Voluntarism is a strategy commonly employed to derail regulation and a well-defined characteristic of neoliberalism. The government's role in ameliorating the social and environmental destruction wrought by capitalism is rolled back, replaced with nonbinding self-regulation (McCarthy and Prudham 2004, 276). Self-determined, self-managed corporate responsibility programs promise that corporate barons will govern in all of our best interests—there is no longer a need for democratically decided limitations and rules on what they do.

There is critical interplay between what is branded as impossible and corporations' utopian pledges of voluntary expert management in all of our best interests. When actual change is outside of the scope of what is considered possible, the best that can be hoped for is some benevolence from the overlords. The intertwining operation of these ideas was marked in the position of the mayor of Kaua'i, who vetoed Bill 2491 on the grounds that (1) the chemical companies should not be antagonized because the island depends on them for jobs and agriculture and (2) regulation is impossible anyway because large corporations are more powerful than local government, so (3) the preferable (and only) option is to work with the industry to find "common ground" through a voluntary solution involving "stakeholders" that will "neutrally" implement the "facts."[††] Chemical companies' triumphs in court have been the vindication of this line of reasoning rather than an indication that wider policy and systemic change is necessary to redistribute power and legalize environmental and health protections.

The idea that we could work together to find "consensus" solutions to the poisoning of land and people is premised on the illusion of common interests. It is a lie of unity between chemical companies and the people they harm in pursuit of profits. This fabrication is intentionally performed globally and locally. DuPont declares: "We face daunting challenges in feeding the world today . . . DuPont and its collaborators has already made great strides in meeting the challenge . . . Because, quite simply, feeding the world is everyone's business" (DuPont 2016). Companies declare their alignment with humanity to "solve hunger" in order to abstract the actual policies, systems, and actors—including themselves—that cause unnecessary hunger in the first place. Locally, companies similarly advertise their oneness with the rest of the community: "We value the island, raise our families here and want to preserve the land and lifestyle for future generations. We're in this together. Kaua'i is our home" (Save Kaua'i Farms 2013). Chemical corporations, workers, and impacted

---

†† See press releases from the mayor's office and newspaper articles in the Garden Island newspaper following the veto of Bill 2491, court ruling, and implementation of the Good Neighbor Program.

residents are presented as occupying the same social positions and interests. Corporations are personified as people, with thoughts, desires, and morals, delinking them from the immoral and impersonal structures to which they are bound.

In the disguising of difference, the operation of power and injustice is masked. Agrochemical corporations most especially proclaim their sameness with an ambiguous "agricultural community," veiling facts of severe exploitation and dispossession that agricultural communities face due to their pursuits of profits and power. In Hawai'i, they lament that local farmers are being pushed out of business by cheap imports, divorcing themselves from the active role they play in constructing such arrangements at the national and international levels. In the dissimulation, they are championed the local saving grace of agriculture for their charity to local farmers. Similarly, migrants displaced from their homes by the capitalist corporate food system, who are now laborers for that system, are represented as unified with the very corporations that have played a role displacing them. In the homogenization of a "local agricultural community," concern, criticism, and questioning of agrochemical-GMO operations is chastised for attacking all farmers, all agriculture, all workers. The ideological construction of unity codes those who speak about power and injustice—about innate divisions in interests—as the cause of division themselves.

Resolving Controversy with Industry Science

In Hawai'i's environmental health controversy, like others, facts and science are allegedly being lost to politics and division. Activists are branded as "anti-science" and accused of intentionally spreading misinformation to stoke controversy. Unity and consensus will purportedly be restored through better education of the public. The subtext of "better education" is uncritical adoption of what the industry (and often state) presents as singular, monolithic, and unchanging science. The reality of uncertainty and incompleteness in scientific knowledge is neglected. Moreover, it is assumed that all facts lead to the same conclusions. In actuality, much divergence in the interpretation of facts revolves around normative, intrinsically political matters. There are always value judgments, different interests, and power struggles at stake, especially in environmental health controversies like Hawai'i's.

Science is not fixed—it is a continual process of exploration, evolution, and contestation. This process is not solely the domain of so-called experts. Following most notably the work of Brian Wynne (1992) around radiation and Chernobyl, social scientists have shown that lay citizens may be better than experts at making room for the unknown and contribute specialisms of their own. Non-experts may engage in rigorous examination and analysis of environmental health through both their experiences and access to official scientific knowledge (Narayan and Scandrett 2014, 564). Despite empirically rich documentation of citizen science and the insights of lay knowledge, communities impacted by polluting industries are seldom considered to be generators of scientific knowledge (Narayan and Scandrett 2014). Instead, those who question the official expertise of state and industry are frequently disparaged as ignorant, hysterical and irrational:

> Who/what is being hurt by the anti-GMO movement, the first casualty is civil discourse, followed by reason and critical thinking. (Conrow)

> Shouldn't we know the type of lunatics we seem to be surrounded by? After all, they're taking over. . . . elected officials are pandering to their endless whining and complaining, policies and laws are being made to address their paranoid fantasies, and in their hysteria, they are ruthlessly going after farms. (KauaiEclectic blog comments 2014)

> Any normal, sane human being would see that there is a clear difference in sanity between the two groups [for/against Bill 2491]." (Testimony in opposition to Bill 2491)

While the agrochemical-seed-biotech industry appeals to notions of homogenous and certain science (there is only science or anti-science), they simultaneously manipulate the fact of always existing scientific uncertainty in order to delay, distort, confuse, and manufacture doubt (Oreskes and Conway 2010). Ideas about science are used in paradoxical ways. They both foreclose what are actually political conflicts between different social interests, and they create the appearance of wide spectrums of scientific debate where there is actually a notable amount of scientific agreement. For example, there is wide scientific consensus on the dangers of certain pesticides, but the fact of always remaining scientific uncertainty and imperfection is exploited by industry to deny culpability. This well-rehearsed corporate tactic of manufacturing doubt is used time and again, in controversies ranging from tobacco to acid rain to climate change (Oreskes and Conway 2010). Thus, the critique here should not be interpreted as minimizing the importance of truth, facts, and science. Indeed, a little more science would be welcome in informing policy debates related to climate change, or in countering corporate deception around their products (Oreskes and Conway 2010). As Naomi Klein (2014) argues in the case of climate change, the fossil fuel industry and right-wing think tanks clearly understand that to accept the dominant scientific consensus would necessitate systemic change. The implications of science have in many ways become politically radical in an era of environmental breakdown, as there is no option but fundamental transformation of systems and dismantling of power that is benefiting from systems and policies as they are. Truly embracing science thus means more politics, not less.

In Hawai'i, the corporate tactic of manufacturing doubt is also executed by conflating concerns over pesticide dangers with concerns around GMOs. Most current regulatory focus in Hawai'i is related to protections from pesticide use, undergirded by a strong body of scientific evidence around harms. While it is true that resistance related to agrochemical operations is often intertwined with a range of social and ecological concerns around GMOs, the dangers of pesticide use are intentionally conflated by industry with murkier scientific debates around genetic engineering. In policy conflicts related strictly to pesticide use, the industry criticizes "ideological bias" against genetic engineering as the motivation of resistance, and then claims "scientific consensus" around the safety of agricultural biotechnology. This serves to distract from core concerns over pesticides and to position the industry on the side of science where they actually have no scientific evidence of pesticide safety to stand on.

Assertions of scientific consensus around the safety and benefits of GMOs are themselves highly partial and misleading. They neglect the complexity of and conflict between scientific views as well as the very different uses to which agricultural biotechnology is being put. As with other environmental issues, scientists themselves disagree to a wide extent on matters of safety and risks of genetic engineering, even when answering seemingly well-defined questions (Carolan 2008; Hilbeck et al. 2015). When the industry, the Alliance for Science, and others brand uncritical promotion of particular technologies as pro-science, they are distorting scientific methodology by not vetting their position against divergent scientific perspectives (AGRA Watch 2020, 12). Moreover, much of the public's concern over agricultural biotechnology lies in matters that exceed narrow risk quantifications, such as normative questions like whether such technologies should exist and to what end they ought to be directed (Carolan 2008, 74; Wynne 2005). While these topics are outside of the scope of this book, the important point here is that resistance to the agrochemical industry based on a broad range of concerns is continuously reframed as anti-scientific rejection of a particular technology. Science is contradictorily cast as the realm that holds answers to social conflicts ranging from whether chemical companies should have to disclose pesticide use, to how and by whom land should be used, to whether patents should be granted on food crops, to whether and how life should be manipulated at the molecular level.

Depoliticizing notions of industry-led and industry-approved science and voluntarism are intertwined with anti-political sentiment and the pervasive notion that there is no alternative to the current order of things. Agrochemical corporations' utopian promises of consensus, harmony, and expert management in "our" best interests grip the public imagination in a wider context of endless realisms that assert the impossibility of meaningful social change. The matrix of these ideologies serves to bolster the current order and make overcoming injustice appear impossible or undesirable. Hawai'i's GMO Ground Zero movement is up against not only an agrochemical-seed-biotech oligopoly and its occupation of the islands, but the root conditions that create the situation and the ideological pillars that buttress it. Resistance here and everywhere must attend most especially to the dual tasks of expanding intersectional systems-oriented organizing and, with it, collective imagination of social possibility.

Chapter 9

Seeds of Possibility

Capitalism has compelled rapid changes in Hawai'i over the past century, from an outpost of US plantation sugar production to a tourism-military monoeconomy with "diversified" agrochemical-GMO operations at the periphery. Recent global maneuvering and consolidation within the agrochemical oligopoly is again changing Hawai'i's landscape. Companies have changed names, reduced operations, and pulled back from some of their most impactful fields, likely in response to growing resistance. As they abandon workers and lands, the voids left will be filled by the next corporate barons if not seized for other purposes. The industry will merely relocate its exploitation, likely to places where protections for people and environment are even more scant and resistance is met with a heavier hand. This has already begun in Puerto Rico.

In Hawai'i and elsewhere, things could be very different from the way they are. Much of the work of movements today is to steadfastly assert this fact, through vision and organization. Swelling waves of intertwined environmental, decolonial, economic, and social justice struggle in Hawai'i hold great potential in this regard. These contemporary struggles are seeded in soils laid by generations prior, including powerful sovereignty, labor, and antiracist movements. They are shaped by Kānaka Maoli epistemologies and practices that have long refused erasure by colonial-capitalist rationalities and intrusions (Goodyear-Ka'ōpua, Hussey and Wright 2014; Silva 2004; Horne 2011). As much as plantations and oligarchies define Hawai'i, so do these glimmers of other worlds.

A Food System for People versus Profit

Hawai'i's movement challenging the agrochemical industry certainly confronts systemic conditions of capitalism. It demands "people over profit" and "health before wealth," and thus speaks to the heart of what capitalism is all about. However, the movement is not generally a radical political struggle. Particular corporations are the primary subject of resistance rather than the logics and processes of capitalism itself. There is a sense that exploitation is the result of aberrant corporations, bad government, or some other form of corrupted capitalism, not the actual core drivers of the system. Agrochemical-seed-biotech company actions are interpreted as the evildoing of moral outliers, versus the rational result of a profit-driven system. The assumption is that good capitalism can exist once government and business is made good—the system itself is rarely up for critique.

Void of systemic critique, Hawai'i's (and other localities') environmental justice conflicts are vulnerable to being twisted into a battle between defense of livelihoods and defense of life and the earth. The inability to counter the pernicious "jobs versus the environment" trap stems partly from lack of structural consideration about economic dependencies, inequalities, and ultimately the structures of capitalism—specifically an oligarchical, plantation-fashioned capitalism. When certain social arrangements are assumed to be natural or unchangeable, it is accurate to say that regulating corporate employers could result in lost jobs, economic deprivation, and paved-over agricultural lands. Livelihoods in competition with the future of life on the planet is an entirely capitalist construct, and one that only intersectional, systemically oriented movements can rescue us from. Even without completely escaping capitalism tomorrow, building a class-based critique and the power of the working class can deliver more equitable and environmentally sane possibilities in the near term. This, however, is not fully centered in Hawai'i's anti-agrochemical industry movement.

Though somewhat short on class and worker based analysis and power building, Hawai'i's anti-agrochemical industry movement is not unconcerned with lost livelihoods. Most typically, local food production is proposed as the straightforward substitute for agrochemical operations. While promising and necessary for many reasons discussed below, there is a need to consider how localization can deliver more liberatory futures when still embedded in the hypercapitalist compulsions that have delivered the corporate industrial food regime. As Raj Patel and Jim Goodman put it, "it's rarely profitable to farm agroecologically when the rules of the game reward ecological devastation, worker exploitation, and monoculture" (2019). The local is celebrated as carrying forth particular values that oppose those of exploitative global capital, but competitive profit maximization still structures local economic relations (Allen 2010; DuPuis, Goodman, and Harrison 2006). For local food businesses restricted by capitalism's structures, profit generation must take at least equal priority to more qualitative and justice-based concerns (DeLind and Bingen 2008). How to make enough to sustain operations and have even a small profit is a real challenge with inflated land values and cheap imports. In other words, the wider context of a corporatized, monopolized, industrialized food system constrains even those attempting to work outside of it.

Furthermore, the structural incentive for exploitation of people and environment still exists at the local level. In Hawaiʻi, a subset of farms producing local food have proven to have worse labor conditions and pesticide practices than agrochemical-GMO operations. In fact, recent pesticide data revealed that just a few large farms producing diversified crops for local consumption and pineapple for export on the island of Oʻahu used nearly two-thirds of all RUPs in recent years. Larger studies elsewhere have also shown that there is nothing inherently less oppressive about small or local for farmworkers (Allen 2010). Of course there are innumerable examples of local farms that prioritize social and environmental goods, but only policy and structural change—not markets—can make these the incentivized norm. Even without an overhaul of the global industrial capitalist food system, policy actions within Hawaiʻi can go far in developing a food system that puts equity and ecological regeneration at the center.

Many Hawaiʻi farmers and food activists are involved in such projects, working beyond markets and consumer demand to change state and county policies in ways that incentivize a more regenerative, healthy, and equitable food system. Examples include policy initiatives around land and water use; publicly funded support for agroecological and traditional Hawaiian agriculture; farm-to-school programs; state procurement of local food; expansion of supplemental nutrition assistance program (SNAP) benefits, including discounted local produce; government-supported distribution systems like farmers markets and food hubs; taxing highly processed imports to fund health and nutrition programs; raising the minimum wage; and more. Local food projects are often layered with anticorporate and sometimes post-capitalist ambition, concerns for equity and the environment, and immediate food justice implications. Localizing food production has the potential to offer significant and prompt environmental benefits and decarbonization; provide livelihoods outside of the service economy and military; deliver nutritious food, especially to those who are food insecure; and remove at least some decision-making from global capital. For geographically isolated islands currently importing an estimated 90 percent of their food, local production can be far more ecologically and economically resilient (Leung and Loke 2008). As the COVID-19 crisis continues, there is widespread focus on just how vulnerable Hawaiʻi's near complete dependence on food imports is and the potential of "a cascade of negative consequences to our food security, health and wellbeing" in a future with an unstable climate (Miles and Merrigan 2020).

Most generally and beyond Hawaiʻi, for those concerned with equity and sustainability through local food, it is critical to assert the impossibility of realizing such goals by way of the very forces that manifest the corporate food regime. A focus on change primarily through markets and raising consumer demand ("vote with your dollar!") results in development of niche ponds of heathy local food for those who can afford it. The wider ocean of corporate-controlled, destructively produced, incredibly unhealthy food remains, and it is not threatened or displaced by niche ponds of "good" food (Allen 2010). The only way to change the larger landscape of food production is through political struggle. Only policy and systems change can transform the food system into one that is actually designed to equitably meet human needs through regenerative agriculture versus one that is designed to maximize profits for the already wealthy.

A critical turn might be marked by merging widespread demands for "people before profit" with goals of local food production in ways that involve more socializing, democratizing, and decolonizing political change. In Hawai'i, rather than opposing capitalist logic as it relates to agrochemical corporations while valorizing local markets as their antithesis, the movement might consider what a local food system by different logics could look like. Again, while changing global capitalism is necessarily a global struggle, local potentials can merge with global ones. To give just a few examples, emphasis could be put on returning stolen lands and waters to Hawaiians; democratizing ownership and workers' cooperatives from production through processing and distribution; living wages for food system workers and all; heavily taxing corporate monopolies that undermine local provisioning (from input suppliers to processed food giants and up to retailers); public procurement to guarantee stable markets for producers; expanding subsidized food for those unable to afford it; open-source science, technology, and seeds; free education in regenerative agriculture; robust incentivization of regenerative agroecological methods, including through strong regulation of negative environmental externalities; public infrastructure for closed-loop waste and nutrient cycling and other infrastructure development for a regenerative local food system. As indicated above, elements of such initiatives are already well underway.

Ultimately, an equitable local food system requires that the people of Hawai'i are able to afford local food. Instead of competing to grow local food as cheaply as possible so working-class people can afford it, the wages of working-class people must rise dramatically while other basic living costs (housing, healthcare, childcare, education, transportation, etc.) decline or are made public and free. Once again, much must be changed at national and international levels. There are constraints to what can be accomplished locally when overarching systems and higher levels of policy are still structured to incentivize a race to exploitative profit generation and capitalist consolidation. However, everything listed here could be at least partially pursued from within Hawai'i while embedded in wider and more globally scaled movements.

Political intervention at the site of local food projects is especially powerful in relation to decolonization and place-based Native Hawaiian initiatives. Historically, political activism has been inseparable from cultural and resource practices, including growing food and taking care of one's land and people (Goodyear-Ka'ōpua, Hussey, and Wright 2014; Gupta 2015). Kanaka food projects often directly challenge and remedy plantation inheritances and colonial dispossessions, including water diversion, knowledge and genetic expropriation, and land restoration. Increasingly, food sovereignty itself is viewed as a pathway toward sustainable self-determination (Gupta 2015). Many 'āina-based projects aim to create livelihoods for young people outside of the tourism-agriculture plantation economy while also dismantling colonial-plantation ideology. One Native Hawaiian educator describes preparing youth for a new future, "helping them to see the environment as their ancestors did and the potential for it to sustain us again." The rising number of creative projects across Hawai'i's indigenous food sovereignty movement affirm and teach a Kānaka view of land, described by Pulama Collier: " 'Āina momona/Fertile land. The land is our ancestor, teacher, parent, provider and nurturer continually shaping and defining us. Hawai'i is an island nation protected, preserved and nurtured by our oceans, lands, sky and heavens. Land/'āina is abundant, rich, and living. We connect to our land as we connect to ourselves. To see our land as 'āina momona is to also see ourselves as full of life, fertile, abundant, and healthy" (Meyer 2014, 100).

As a decolonizing project, Hawaiian food sovereignty is connected to the wider food sovereignty movement, led especially from the Global South. More than market- and consumer-centric efforts to increase local food, food sovereignty situates the development of regional food systems within the politics of wealth and resource redistribution, producers' rights to the means of production (including indigenous rights to their land, water, and seed), democratization, and dismantling transnational agribusinesses' monopoly power (Holt-Gimenez and Shattuck 2011; Via Campesina 1996). Ultimately, the food sovereignty movement is a vision for a food system that is about meeting the needs of people and the earth versus the profit desires of the one percent.

Decolonization and Settlers in the Occupied State

A significant political and cultural renaissance began in Hawai'i in the late 1960s and 1970s, and today it has escalated to a broad decolonization movement encompassing diverse tactics and aims. The strength of the movement is evident in water rights wins, protection of sacred places, land occupation and reclamation, legal fights for political independence, uncovering and retelling of history, language and cultural revival, Hawaiian charter schools, and much more. Kamanamaikalani Beamer writes that Native Hawaiians "have endured successive attacks on our national identity, our lands, and our language . . . Through the reclamation of our collective past and a surge of political and cultural awareness, the national consciousness has reemerged" (2014, loc. 3507 of 5334). The vibrancy, intelligence, and growing power of the movement is a force that is shaping the future of the islands and indicates the potential—through struggle—of more just, equitable, and emancipatory horizons.

Because the agrochemical industry's occupation of Hawai'i brings to surface the long and continuing history of the imbrication of plantation agriculture and imperial occupation, decolonization and agrochemical industry resistance frequently converge. Place-based indigenous food practices, water rights battles between Hawaiian taro farmers and chemical companies (Henkin and Moriwake 2013), the leasing of "state" (seized) lands to agrochemical operations, biopiracy and intellectual property patents (Kanehe 2014), and demands for political and resource self-determination (Goodyear-Ka'ōpua, Hussey, and Wright 2014) are just some of the sites where this is most visible. Resistance to agrochemical operations has also been salient in widening aloha 'āina movements. Though most typically translated simply as "love for land," the concept of aloha 'āina is more capaciously about spiritual belief, land management ethic, and Hawaiian sovereignty (Silva 2004; Beamer 2013). Contrary to how it is used by the chemical companies and economic-political elite more generally to quiet and shame dissent, "aloha" itself "requires one to speak in the face of injustice" (Beamer 2013, 13). Pushing back against ways in which "aloha" has long been used in attempts to depoliticize and individualize issues of structural injustice, Kamanamaikalani Beamer writes: "Aloha is active and something that needs to be put into practice, not something that is a state of being. The problems around social, cultural, and ecological justice in Hawai'i are not insignificant, nor are they something that we can will away through selfless compassion" (2013, 13).

To invoke aloha is to call for actively confronting systemic injustice, and to organize around aloha 'āina is to link social, cultural, and ecological justice (Beamer 2013, 11). Aloha 'āina encompasses belief of the reciprocal relationship between people and 'āina, and responsibility to protect 'āina. Distinct from Euro-American concepts of nature as separate from human culture, aloha 'āina is about interconnection and human embeddedness in the wider web of life. It has been a rallying call for Native Hawaiian rights and sovereignty, including the 1970s movement to protect the island of Kahoʻolawe from military bombing and the cultural revival of the following decades (McGregor 2002). In recent years aloha ʻāina has united a multitude of struggles around land, food and agriculture, development and urbanization, protection of sacred places, Native Hawaiian rights, and sovereignty. Veteran Kanaka activist Walter Ritte asserts that "if we don't come together, if we don't unify, we are going to lose [our natural resources]. . . . We can conquer everything if we can come together . . . The banner that puts all of us in the same room, under the same coalition, is aloha ʻāina. If we all love the land, we can be on the same team, no matter what nationality you're from . . . That's the end goal, to build our coalition under the banner of aloha ʻāina" (2015).

While decolonization and resistance to agrochemical operations are intimately connected, they are not always in straightforward or synergistic relationship. Aloha ʻāina, for instance, can be a slogan that many mobilize behind without understanding or supporting its decolonial significance. There is often objection to the agrochemical industry without concern for United States occupation and the dispossession of Hawaiians as the most fundamental condition of the oligopoly's functioning. The GMO Ground Zero movement must center and directly confront this fact if it is to address the systemic causes and effects of the agrochemical industry in Hawaiʻi. This includes accounting for the role of settlers in the movement, who, for example, can also be entangled in processes of gentrification and displacement of Native Hawaiian and working-class people from the islands. Related, "the oppressive role white cultural imperialism has played and continues to play in Hawaiʻi" is generally neglected by white settlers (Trask 1991, 1212). White activists sometimes do not recognize long histories of racial hierarchy and violence in the islands, and how their own discourses energize rather than dismantle racism. For instance, comments about "plantation mentality" disparage the agency of working-class people of color and erase long histories of radical resistance by "plantation people" (Horne 2011; Beechert 1985). As Chris Leong describes, "condescending" and "patronizing attitudes" have been woven in with structures of racial oppression since white imperialists arrived in the islands (1997). Reminiscent tones of racialized superiority are experienced today when white settlers victimize or instruct those living near agrochemical fields. Racialized savior mentality stands in the way of solidarity and intersectional movement building. Many settler activists are aware of these dynamics and determinedly organizing within the movement to educate, build more intersectional solidarity, and dismantle internal oppression. Like all highly decentralized movements, less-than-liberatory elements always need to be challenged, and there is concentrated work within the movement to do so.

While not without contradiction, Hawai'i's anti-agrochemical industry movement is highly entwined with numerous nodes of resistance to US occupation and imperialism. Given the significance of large capitalist agribusiness to Hawai'i's continuing colonial history and power structure, its centrality in recent struggles is notable. Moreover, there are convergences of movements that indicate deeper solidarity and commitment to wider change. It is in this intersectionality of struggle—as witnessed in recent aloha 'āina mobilizations—that single-issue-based activism might transform into more radical systemic consciousness and movement building at the root (Davis 2016).

Internationalism

Focus on the particularities of the issue and local situation is critical to immediate and urgent environmental justice. At the same time, the movement's wider emancipatory potentials lie in simultaneously advancing systemic and global analysis and solidarity. The global political economy shapes the local landscape in ways that must be understood and countered. Ultimately, the globalized nature of racialized capitalist exploitation and its entwinement with imperialism demands internationalism—which can be locally rooted—to match it.

As might be noted more generally in social movement struggle today, capitalism's globalism is frequently thought to be best countered through "the local" rather than through a global-local politics that also claims a universal ethic. When internationalism is abandoned, even in unintended ways, localism can work to narrow the frames of care and responsibility. For instance, scholars of agrifood movements have identified the rise of a type of defensive localism that encourages responsibility and care within demarcated boundaries and pits regions against one another (Hinrichs 2003; see also Allen 2010; Harvey 1996; Allen, FitzSimmons, et al. 2003; DuPuis and Goodman 2005). Given capitalism's competitive and divisive dynamics, these tendencies are not unexpected. Their appeal in social movements indicates openings for reactionary politics. A challenge for Hawai'i's movement is to continually seek ways to change the local situation that simultaneously involve advancing global justice and solidarity.

While there is need to be cautious of defensive localism tendencies, it would be a mistake to characterize even highly localist engagements in Hawai'i as necessarily lacking internationalism. The movement in Hawai'i is a product and an aspect of the constant global circulation of ideas, strategies, experiences, and relationships. Local and global actions feed one another. There is ever increasing connection and collaboration among activists worldwide. Especially in proliferating food sovereignty, energy and climate, mining, and local environmental and resource struggles, sites of resistance are frequently globally linked in praxis (Klein 2014; Via Campesina 1996). Moreover, internationalism and system-oriented struggle can emerge from within resistance that is initially compelled by what is proximate and immediate. In Hawai'i, connections with international food sovereignty and justice movements have expanded local activists' analysis and focus. Internationalization has opened space for more radical critique as Hawai'i's movement interacts with struggles of the Global South. Peasant farmers and farmworkers, especially women and indigenous people, have traveled to Hawai'i to teach, learn, and build solidarity. They have informed the movement not only about particular agrochemical injustices in other locales, but also about related matters of US imperialism and trade, gender oppression, poverty, inequality, migration, and agricultural workers.

People's engagement at the local level is often in part motivated by desire to affect the global. What happens in Hawai'i has repercussions far beyond the islands. Today's globally integrated supply chains create points of extreme vulnerability for capital. Hawai'i is a strategic node of resistance to the agrochemical industry's global production and power. Many activists are neither oblivious nor ambivalent to this. Even highly localized engagements can be undergirded by goals of global justice. A most important matter becomes how to take cross-border solidarity to a level beyond knowing and caring about the plight and fight of others to articulating common cause and coordinated struggle.

This challenge is most salient in regard to Puerto Rico. Also a US island colony with three or four growing seasons per year, Puerto Rico has historically been second only to Hawai'i in the number of genetically engineered field trials conducted by the agrochemical industry. As resistance in Hawai'i has grown, companies are expanding their presence in Puerto Rico. Over the past decade, substantial public lands and financial resources have been quietly transferred to the industry at the same time as imposed debt and austerity rob Puerto Ricans of basic necessities. Resistance within Puerto Rico is starting to boil as people living near fields express similar concerns about pesticide use. The details of how activists in Hawai'i organize alongside Puerto Ricans as the industry relocates to evade regulation and resistance are still evolving. Many activists in Hawai'i understand that isolated, singularly focused struggle in a world of globalized capitalism can equate to simply chasing Monsanto out of one's own backyard. Ultimately, it will take deeper levels of globally connected organizing to sustain the level of challenge necessary to overcome consolidated global capitalist power.

Influence on and by the Mainstream American Food Movement

In bringing pesticides, producing communities, and collective political engagement to the forefront, Hawai'i's movement works to shift a conversation that, in recent years, has been somewhat dominated in the United States by individualized consumer concerns and market-based action (Guthman 2008). Mainstream food activism in the US has tended to reproduce neoliberalism's core ideas of social change through individual behaviors, markets, and entrepreneurialism. Such trends indicate a moment marked by depoliticization and lack of critical thought about or beyond the structural compulsions of capitalism (Guthman 2008; DuPuis and Goodman 2005; Allen, FitzSimmons, et al. 2003; Alkon and Agyeman 2011; Alkon and McCullen 2011). Capitalist agriculture's imperatives to accumulate, consolidate, dispossess, and exploit appear to be ever expansive and are hardly threatened by a rapid rise in organic and fair trade, farmers markets, and public attention to food matters. While niche markets of good food for those who can pay are increasingly available, rapidly destructive agricultural production also grows. Corporate agriculture has even gained new profit terrains in recuperation of social ideals as Walmart goes "local," Nestle goes "fair," and organic dollars go to Coca-Cola, General Mills, and other food-processing giants (Guthman 2004; Fridell 2007; Jaffee and Howard 2010; Hauter 2012).

The general discourse around and critique of genetically modified foods can similarly be individualized, market focused, and void of structural critique. While activist concerns related to genetic engineering are broad, they are often framed as problems of the thing itself (e.g., anti-GMO) rather than problems of the social conditions, relations, and paradigms that make a technology function as it does. This is not to dismiss that many are concerned specifically with the potential consequences of manipulating organisms' genes. Rather, the point is that considerations of—and possibilities for—systemic change remain limited when the core subject of critique is technical, not social. Capitalist enclosure, commodification, exploitation, and power come to look like problems of a technology rather than the basis from which particular uses of technology materialize. In the most limited of anti-GMO activism, the endpoint might be an agricultural system purged of genetically engineered foods but thick with patents on conventionally bred plants, controlled by a few mega companies, and continuing to incentivize highly environmentally destructive practices and technologies in the pursuit of profit.

While mainstream US activism around GMOs has centered largely on consumer health, individuals' right to choose, and corresponding campaigns to label GMO-containing foods, Hawai'i has begun to bring to the forefront things that have been somewhat sidelined. First, it makes clearer the ways in which a handful of colluding corporations are using biotechnology to entrench a pesticide-intensive, industrial-style agricultural system in which they control both the seeds and the chemicals. Second, it draws attention to producer communities and environmental justice concerns. And third, it illustrates the need for and inspires the possibility of collective political action. These are critical openings, as they orient a more systemic critique and attention to issues of justice while also challenging neoliberal rationality that justice can be achieved through the market.

Moreover, while sometimes lacking in structural critique, it is also true that concerns about biotechnology are frequently tied to demands for the commons. Most notably, social movements are rejecting the idea of seeds, genes, and life as ownable. In Hawai'i's movement, these calls for the commons extend to land, water, and air as elements of existence that all of humanity must be stewards of. Often grounded in Hawaiian epistemology, the movement declares that people have a collective kuleana (responsibility and privilege) to mālama (take care of) these commons for both future generations and the nonhuman world. Commitment to the commons undergirds objection to the agrochemical industry's use of land and informs alternative visions of using land in ways that serve the collective good. These ideas indicate critical yearnings and struggles for change at the root, regardless of whether they are explicitly anticapitalist.

Seizing the Possible

The greatest potentialities of Hawai'i's anti-agrochemical industry movement lie in inspiring and influencing wider systemic struggle, both organizationally and ideologically. This is not to abandon the specificities of the issue, but to think and organize in ways that are also systemic and intersectional. Rather than an outside prescription, this is an observation of much that is already happening: the development of solidarity across issues that are joining in the streets, advocating and organizing together; policy changes that directly confront Hawai'i's plantation history and offer protections to those who have long suffered its consequences; and the invoking of values and possibilities that upset capitalist and colonial rationality.

A most critical battleground of all social movements today is over what is considered possible in the social order. Neoliberalism's colonization of the collective imagination has left activists and those to whom they seek to appeal often incapable of thinking outside of the limitations of not merely capitalism, but capitalism in its most rapacious form. It is essential to identify, expand, and embolden the liberatory threads within social justice struggles, as they contain seeds of freedom from neoliberalism's ideological grip. In Hawai'i's anti-agrochemical industry movement, demands for "health before wealth" and "people before profit" call attention to capital's core drives and invoke antithetical values. There is articulation of the fundamental conflict between the logic of capitalism and the very things most would claim to value. This includes the movement's call for the commons—for the recognition that some things belong to us all and that it is humanity's collective responsibility to steward the earth. Capitalism's core drive to privatize what has and could be common is a direct point of resistance, especially in regard to seeds and genes.

Further, the movement envisions and seeks to create systems of production structured by logics different from capitalism. Specifically, there is an attempt to embed food production in values of ecological sustainability and equitable access to nutritious food. Building alternative food systems is also viewed as a way to escape chains of global corporate exploitation. While structurally limited, what people are aspiring for is a production system that is about meeting human and environmental needs rather than about the ever-expansive accumulation of profit. Again, this is a critical opening that could move in more system-oriented directions of socializing, democratizing, and decolonizing food production and distribution. There has long been a relationship between land, food, and cultural and political sovereignty in Hawai'i, and the movement feeds these struggles as it exposes and challenges the intimate connection between colonialism and capitalist agribusiness.

In all of these elements, Hawai'i's agrochemical struggles incite both collective responsibility and power, a direct assault on neoliberalism's depoliticizing and individualizing aims. Social change is not left to the terrain of market-based activism for those who can afford to vote with their dollar. Rather than an individualized "not in my body" politics of shopping decisions, Hawai'i's struggle against the agrochemical-GMO industry is about fundamental matters of communities most impacted by production, and changing policy and systems through collective action. In activists' bold challenge to global chemical companies, they have inspired belief in—and illustrated the necessity of—social change through bottom-up people's power.

In these ways and others, Hawai'i's movement begins to unshackle collective imagination from capitalist and colonial logics. These are critical openings that can be expanded. Movement work includes not just practical outcomes, but constant "revolutions at the level of common sense" (Graeber 2012, 165).

Even from within the specificities of a political conflict, the wider horizon of what is considered to be possible in the social order must constantly be pushed. If today's "common sense" of social possibility is left unchallenged, aspirations for justice are severely diminished. Godfather of neoliberalism Friedrich Hayek understood this well, describing it the "war of ideas": "Those who have concerned themselves exclusively with what seemed practicable in the existing state of opinion have constantly found that even this has rapidly become politically impossible as the result of changes in a public opinion which they have done nothing to guide" (Hayek 1949, 384). The corporate countermovement in Hawai'i provides valuable insight into how possibility is narrowed and disciplined, and why the grander war of ideas is a most critical terrain of all struggle. This battle is only more intense at larger scales and as progressive movements gain power. Elites and their ubiquitous mechanisms of propaganda are working ceaselessly to lock down the future by convincing the masses that anything more humane is an impossibility. Whether in the fight for climate justice, abolition of student debt, defunding the police, nationalizing banks, pesticide protections, clean water, taxing the rich, or overthrowing capitalism, all political battle today involves recoding what is believed to be attainable—recovering possibility from the terror of ideologically imposed "impossibility."

When the ideological instruments of impossibility—of "no alternative" or a "natural" order of things—are removed, what we are left with is a society in which the twenty-six richest people are as wealthy as the poorest half of the world (Oxfam 2019). Alongside unfathomable inequality and unnecessary human suffering, we have a system that is compulsively producing the non-negotiable planetary conditions for its own extermination. The urgent gravity of the global situation necessitates radical—at the root—response. There is no future in capitalism's "realisms"—in the bleak myth that we must choose between our livelihoods or the earth, the economy or the environment, people or the planet. Even dwarfed bits of justice are said to be impossible when we limit our demands and struggles to fit within these confines of the present. When wider systems change is not an option, circumstances appear immutable and injustice is enshrined as the inevitable order of things. This is not to suggest that justice is achievable only with the complete annihilation of capitalism and its interdependent forces of racism and imperialism. Gains can be made that bring us toward more emancipatory horizons from within the conditions of the present. Harnessing immanent and immediately available possibilities has the potential to open new horizons of possibility and unbind constrictions on the thinkable. Thus, even more limited and reformist struggles must engage in the wider war of ideas, denaturalizing the cruelty of this age and feeding the potentials of radically different, positively more humane futures.

Without doubt, this is a historic moment of decision. Fascism, xenophobia, more extreme austerity, neoliberalism, and fattened empire are all scrambling to claim power from multiple intersecting crises. Signified in the elections of Donald Trump, Rodrigo Duterte, Jair Bolsonaro, and others, fascism has returned from the margins to the mainstream. As Naomi Klein and others have warned, fascism is and will continue growing with climate change. We are already in the dawn of "climate barbarism" (Hanman 2019). At the same time, people's movements are growing more intersectional and systemically oriented, both in ideology and organization. They are asserting bold visions that actually take seriously the fact that we must change everything in order to avoid catastrophic climate unraveling. Even in the United States, there are signs that the war of ideas is beginning to be won by strongly progressive forces. Polls show a majority of Americans consistently favoring the Green New Deal, universal health care, increased taxes on the rich, tuition-free college, paid parental leave, raising the minimum wage, and other measures that defy the decades-long neoliberal consensus. A majority of young people go as far as to say that they do not support capitalism. The battle over public common sense is raging as neoliberalism's ideological lockdown is collapsing. This is only ramping up as COVID-19 unveils the remarkable racial, economic, gender, and environmental injustices of the present.

While the pace of change is quickening, the direction of that change is always up for grabs. The scale of injustice and brutality in our present is often used to provoke despair and cynicism—to blunt resistance by convincing us that humans are incapable of anything better. Today's movements must recognize and politicize the fact that we are capable of so much more. Despite the inhumanity that our system demands, we do not abandon our most intrinsic human drives for mutual aid, compassion, and solidarity. We are "supreme cooperators" with an astonishing degree of altruism, and we evolved to be that way (Monbiot 2017, 14). This is most evident, all around us, all the time, in people's moment-to-moment generosity and sensitivity to the well-being of others. Though frequently negated, overlooked, suppressed, or enclosed, such capacities could be the driving force of our public lives together. We must not end with the tale of futility, taking today's horrors as evidence that we are incapable of something better. The world that the vast majority of us long for is actually here, innate in our most common interactions and aspirations. It only needs to be freed.

Bibliography

Abercrombie, N. 2013. "Governor Works with Kauai Legislators to Address Community Concerns on Pesticides." Press Release, State of Hawaii Governor's Office. 23 September. https://votesmart.org/public-statement/811620/governor-works-with-kauai-legislators-to-address-community-concerns-on-pesticides#.VxwW6T9FV3t.

ABSA (American Biological Safety Association). 2015. Letter submitted to US National Science and Technology Council, Re: Docket No. FDA-2015-N-3403. 12 November. https://www.absa.org/pdf/151112ABSACoordinatedFrameworkComments.pdf.

Achitoff, P., Kimbrell, G., and Wu, S. 2015. INTERVENOR-DEFENDANTS-APPELLANTS' OPENING BRIEF in SYNGENTA SEEDS, INC., *et al.* Plaintiffs-Appellees, v. COUNTY OF KAUA'I, Defendant-Appellant and KA MAKANI HO'OPONO, *et al.* Case No. 1:14-cv-00014-BMK.

Adair, S. 2010. "The Commodification of Information and Social Inequality." *Critical Sociology*, 36(2): 243–263.

AGRA Watch. 2020. *Messengers of Gates' Agenda: A Case Study of the Cornell Alliance for Science Global Leadership Fellows Program.* Seattle: AGRA Watch.

Alkon, A., and Agyeman, J. 2011. *Cultivating Food Justice: Race, Class, and Sustainability.* Cambridge, MA: MIT Press.

Alkon, A., and McCullen, C. 2011. "Whiteness and Farmers Markets: Performances, Perpetuations . . . Contestations?" *Antipode*, 43(4): 937–959.

Allen, P. 2010. Realizing Justice in Local Food Systems. *Cambridge Journal of Regions, Economy and Society*, 3(2): 295–308.

Allen, P., FitzSimmons, M., Goodman, M., and Warner, K. 2003. "Shifting Plates in the Agrifood Landscape: The Tectonics of Alternative Agrifood Initiatives in California." *Journal of Rural Studies*, 19(1): 61–75.

Allen, R., Gottlieb, M., Clute, E., Pongsiri, M., Sherman, J., and Obrams, G. 1997. "Breast Cancer and Pesticides in Hawaii: The Need for Further Study." *Environmental Health Perspectives*, 105(3): 679–683.

Altemus-Williams, I. 2013. "The Struggle to Reclaim Paradise." *Waging Nonviolence.* 10 April. https://wagingnonviolence.org/2013/04/the-struggle-to-reclaim-paradise.

Altonn, H. 1998. "Biotechnology: Our Next Industry?" *Star-Bulletin.* 26 February. http://archives.starbulletin.com/98/02/26/news/story5.html.

"Alyssa Katz on the Influence Machine." 2015. *Corporate Crime Reporter*, 33(12). 31 August. http://www.corporatecrimereporter.com/news/200/alyssa-katz-on

-the-influence-machine-the-chamber-of-commerce-and-the-corporate-capture-of-american-life.
Aole GMO Means Aole Aloha, n.d. https://nogmomeansnoaloha.com/.
Amin, S. 2009. *Eurocentrism,* 2nd ed. Translated by Russell Moore and James Membrez. New York: Monthly Review Press.
Andrade, C. 2008. *Hā'ena: Through the Eyes of the Ancestors.* Honolulu: University of Hawai'i Press.
Aoudé, I. 2001. "Policy of Globalization and Globalization of Policy." *Social Process in Hawaii,* 40: xi–xxvii.
Argyres, N., and Liebeskind, J. 1998. "Privatizing the Intellectual Commons: Universities and the Commercialization of Biotechnology." *Journal of Economic Behavior & Organization*, 35(4): 427–454.
"Asked to Suggest a Policy." 1898. *Pacific Commercial Advertiser.* 10 March. Accessed through *Chronicling America: Historic American Newspapers.* Library of Congress.
Azmanova, A. 2015. "The Crisis of the Crisis of Capitalism." Unpublished paper. University of Kent. https://www.academia.edu/11034347/The_Crisis_of_the_Crisis_of_Capitalism.
Bacon, D. 1995. "Trouble in Paradise: Hawaii's Sugar Workers Fight for a New Life." *David Bacon Stories,* 24 September. http://dbacon.igc.org/Work/01Sugar.html.
Badgley, C., Moghtader, J., Quintero, E., Zakem, E., Chappell, M., Aviles-Vazquez, K., Samulo A., and Perfecto, I. 2007. "Organic Agriculture and the Global Food Supply." *Renewable Agriculture and Food Systems*, 22(2): 86–108.
Baker, R. S. 1911. "How King Sugar Rules in Hawaii." *Wonderful Hawaii: A World Experiment Station.* Printed by *The American Magazine.* 28–38.
Banner, S. 2005. "Preparing to Be Colonized: Land Tenure and Legal Strategy in Nineteenth-Century Hawaii." *Law & Society Review,* 39(2): 273–314.
Barnett-Rose, R. 2015. "Judicially Modified Democracy: Court and State Preemption of Local GMO Regulation in Hawai'i and Beyond." *Duke Environmental Law and Policy Forum*, 26: 71–130.
Beamer, K. 2013. "Tutu's Aloha 'Āina Grace." In Goodyear-Ka'ōpua, N. (Ed.), *The Value of Hawai'i 2: Ancestral Roots, Oceanic Visions.* Honolulu: University of Hawai'i Press.
Beamer, K. 2014. *No Mākou ka Mana: Liberating the Nation.* Honolulu: Kamehameha Publishing.
Beamer, K., and Duarte, T. 2009. "I Palapala No Ia Aina–Documenting the Hawaiian Kingdom: A Colonial Venture?" *Journal of Historical Geography*, 35(1): 66–86.
Beamer, K., and Tong, W. 2016. "The Māhele Did What? Native Interest Remains." *Hūlili: Multidisciplinary Research on Hawaiian Well-Being,* 10: 125–145.
Beechert, E. 1985. *Working in Hawaii: A Labor History.* Honolulu: University of Hawai'i Press.
Benbrook, C. 2001. "Do GM Crops Mean Less Pesticide Use?" *Pesticide Outlook*, 12(5): 204–207.
Benbrook, C. 2012. "Impacts of Genetically Engineered Crops on Pesticide Use in the US—The First Sixteen Years." *Environmental Sciences Europe*, 24(24): 1–13.
Berg, C. 2016. "Pesticide Screening on Kauai." Comments submitted to Kauai Joint Fact Finding Group. 8 April.
Berman, R. 2014. "Win Ugly." Presentation to Western Energy Alliance. Colorado Springs, 25 June. https://www.youtube.com/watch?v=vsgGow4onK8.

" 'Big 6' Pesticide and GMO Corporations." 2015. SourceWatch. http://www.sourcewatch.org/index.php/%22Big_6%22_Biotech_Corporations.

"Big Five." n.d. HawaiiHistory.org. http://www.hawaiihistory.org/index.cfm?fuseaction=ig.page&PageID=29.

"The Biologist Who Challenged Agribusiness." 2016. *New Yorker.* 1 December. https://www.newyorker.com/culture/culture-desk/the-biologist-who-challenged-agribusiness.

"Biotechs Call Cayetano Their Governor of the Year." 2002. *Pacific Business News*, 22 October. https://www.bizjournals.com/pacific/stories/2002/10/21/daily22.html.

Bisaillon, L. 2012. "An Analytic Glossary to Social Inquiry Using Institutional and Political Activist Ethnography." *International Journal of Qualitative Methods*, 11(5): 607–627.

Bishop, A. 1838. "An Inquiry into the Causes of Decrease in the Population of the Sandwich Islands." *The Hawaiian Spectator*, 1(1). Conducted by An Association of Gentlemen. Honolulu: Edwin O. Hall Printer. 52–66.

Block, F. 2008. "Swimming against the Current: The Rise of a Hidden Developmental State in the United States." *Politics & Society*, 36(2): 169–206.

Bond-Graham, D. 2013. "Monsanto University." *CounterPunch.* 27 February. http://www.counterpunch.org/2013/02/27/monsanto-university.

Bouchard, M., Bellinger, D., Wright, R., and Weisskopf, M. 2010. "Attention-Deficit/Hyperactivity Disorder and Urinary Metabolites of Organophosphate Pesticides." *Pediatrics*, 125(6): e1270–e1277.

Bouchard, M., Chevrier, J., Harley, K., Kogut, K., Vedar, M., Calderon, N.,Trujilo, C., Johnson, C., Bradman, A., Barr, D., and Eskenazi, B. 2011. "Prenatal Exposure to Organophosphate Pesticides and IQ in 7-Year-Old Children." *Environmental Health Perspective*, 119(8): 1189–1195.

Bowring, F. 2003. "Manufacturing Scarcity: Food Biotechnology and the Life Sciences Industry." *Capital and Class*, 27(1): 107–144.

Brewbaker J. 2003. *Corn Production in the Tropics: The Hawai'i Experience*. University of Hawai'i Manoa, College of Tropical Agriculture and Human Resources.

Brown, W. 2007. "Democracy and Bad Dreams." *Theory and Event*, 10(1).

Brown, W. 2015. *Undoing the Demos: Neoliberalism's Stealth Revolution*. Cambridge, MA: MIT Press.

Brown, W., and Shenk, T. 2015. "Booked #3: What Exactly Is Neoliberalism?" *Dissent*. 2 April. https://www.dissentmagazine.org/blog/booked-3-what-exactly-is-neoliberalism-wendy-brown-undoing-the-demos.

Bryant, H., Maisashvili, A., Outlaw, J., and Richardson, J. 2016. *Effects of Proposed Mergers and Acquisitions among Biotechnology Firms on Seed Prices*. Working Paper 16-2, Agricultural and Food Policy Center, Texas A&M University.

Bunge, J. 2015. "Monsanto to Chart Growth Plan as Farmers Feel Squeezed." *Wall Street Journal*. 4 October. http://www.wsj.com/articles/monsanto-to-chart-growth-plan-as-farmers-feel-squeezed-1443864782.

Bynum, T. 2013. "2491 a Chance to Control Destiny." *Garden Island*. 29 September. http://thegardenisland.com/news/opinion/guest/a-chance-to-control-destiny/article_1f3a3742-28d4-11e3-960c-0019bb2963f4.html.

Campbell, E., Louis, K., and Blumenthal, D. 1998. "Looking a Gift Horse in the Mouth." *Journal of the American Medical Association*, 279(13): 995–999.

Conrow, J. Ongoing. *KauaiEclectic Blog*. http://kauaieclectic.blogspot.com.

Carolan, M. 2008. "The Multidimensionality of Environmental Problems: The GMO Controversy and the Limits of Scientific Materialism." *Environmental Values*, 17(1): 67–82.

Caron, W. 2015. "UHM Faculty Condemn Academic Freedom Violations." *Hawai'i Independent*. July 21. http://hawaiiindependent.net/story/uhm-faculty-rally-condemn-academic-freedom-violations.

Castells, M. 2005. *The Power Of Identity*. Malden, MA: Blackwell.

Castle, S. 1869. "Mr. Castle and the Labor System." *The Pacific Commercial Advertiser*. 24 April. Accessed through *Chronicling America: Historic American Newspapers*. Library of Congress.

Cayetano, B. 2000. State of the State Address. 24 January. http://archives.starbulletin.com/2000/01/24/news/story1a.html.

CEDS. 2010. *Hawaii Statewide Comprehensive Economic Development Strategy*. Prepared by the State of Hawaii, Office of Planning of Department of Business Economic Development and Tourism. http://files.hawaii.gov/dbedt/op/spb/Final_CEDS_2010.pdf.

Center for Food Safety (CFS) and Save Our Seeds (SOS). 2013. *Seed Giants Versus U.S. Farmers*. http://www.centerforfoodsafety.org/reports/1770/seed-giants-vs-us-farmers.

Chamber of Commerce Hawaii. N.d. "Business Advocacy." https://www.cochawaii.org/business-advocacy-hawaii.

"The Chamber of Secrets." 2012. *The Economist*. Print edition, business section. 21 April. http://www.economist.com/node/21553020.

Charles, D. 2001. *Lords of the Harvest: Biotech, Big Money, and the Future of Food*. Cambridge, MA: Perseus.

Chopra, T. 2015. "Persistent Narratives, Persistent Failures: Why GM Crops Do Not—and Will Not—'Feed the World.'" *Canadian Food Studies/La Revue canadienne des études sur l'alimentation*, 2(2): 209–216.

Cocke, S. 2013a. "Pesticide Contamination of Drinking Water Worries State Health Officials." *Civil Beat*. 8 March. http://www.civilbeat.com/2013/03/18548-pesticide-contamination-of-drinking-water-worries-state-health-officials.

Cocke, S. 2013b. "Gov Wades into Hawaii Pesticide Debate as Kauai Poised to Take Action." *Civil Beat*. 25 September. http://www.civilbeat.com/2013/09/19986-gov-wades-into-hawaii-pesticide-debate-as-kauai-poised-to-take-action.

Cocke, S. 2013c. "Does Hawaii's Failure to Enforce Pesticide Use Justify Action by Kauai?" *Civil Beat*. 5 October. http://www.civilbeat.com/2013/10/20066-does-hawaiis-failure-to-enforce-pesticide-use-justify-kauais-action.

Coffman, T. 1998. *Nation Within: The Story of America's Annexation of the Nation of Hawaii*. Fairbanks, AK: Epicenter.

Colby, G. 1984. *DuPont Dynasty: Behind the Nylon Curtain*. New York: Open Road.

Cooper, G., and Daws, G. 1985. *Land and Power in Hawaii: The Democratic Years*. Honolulu: University of Hawai'i Press.

Cruz, K. 2019. "Federal Judge Finds Hawai'i Agribusiness Development Corporation Is Unlawfully Contaminating Kaua'i's Shores." EarthJustice Blog, 10 July. https://earthjustice.org/news/press/2019/federal-judge-finds-hawai-i-agribusiness-development-corporation-is-unlawfully-contaminating-kaua-i-s-shores.

CTAHR (College of Tropical Agriculture and Human Resources). N.d. "Biotechnology and Agricultural Education Program: CTAHR Research." http://www.ctahr.hawaii.edu/biotech/CTAHR_Research.html.

Curtis, H. 2014. "Launching a High Technology Industry in Hawaii." *Ililani Media*. 19 May. http://ililanimedia.blogspot.co.nz/2014/05/launching-high-technology-industry-in.html.

D'Angelo, C. 2013. "Mana March Draws Thousands." *Garden Island*. 9 September. https://www.thegardenisland.com/2013/09/09/hawaii-news/mana-march-draws-thousands.

D'Angelo, C. 2014. "The Tip of the Iceberg." *Garden Island*. 7 October. http://thegardenisland.com/news/local/the-tip-of-the-iceberg/article_2ede7b44-4deb-11e4-8cb7-e33aa52f0a8c.html.

Darby, G., and Jussawalla, M. 1993. "Telecommunications." In Roth, R. (Ed.) *The Price of Paradise*. Vol. II. Honolulu: Mutual Publishing. 45–50.

Davis, A. 2016. *Freedom Is a Constant Struggle: Ferguson, Palestine, and the Foundations of a Movement*. Chicago: Haymarket Books.

DBEDT (Department of Business, Economic Development, and Tourism), State of Hawaii. 2007. *Hawaii Enterprise Zones Partnership Program: Report to the Governor for Calendar Year 2007*. Strategic Marketing and Support Division. Honolulu.

DBEDT (Department of Business, Economic Development, and Tourism), State of Hawaii. 2012a. *Hawaii Enterprise Zones Partnership Program Annual Report CY 2010–2012*. Honolulu.

DBEDT (Department of Business, Economic Development and Tourism), State of Hawaii. 2012b. *Increased Food Security and Food Self-Sufficiency Strategy, Volume II: A History of Agriculture in Hawaii and Technical Reference Document*. Prepared in cooperation with the Hawaiʻi Department of Agriculture. Honolulu.

De Lama, G. 1994. "Sugar King No More in Hawaii." *Chicago Tribune*, 19 June. http://articles.chicagotribune.com/1994-06-19/business/9406190289_1_sugar-mill-sugar-planters-sugar-workers.

DeLind, L., and Bingen, J. 2008. "Place and Civic Culture: Re-thinking the Context for Local Agriculture." *Journal of Agricultural and Environmental Ethics*, 21(2): 127–151.

DePledge, D. 2014. "Seed Money." *Star-Advertiser*. 13 January. http://politicalradar.staradvertiserblogs.com/2014/01/13/seed-money.

De Schutter, O. 2010. *Report to the UN Human Rights Council by the Special Rapporteur on the Right to Food*. New York: UN General Assembly.

De Schutter, O. 2011a. *Agroecology and the Right to Food*. Report presented at the 16th Session of the United Nations Human Rights Council. Geneva, Switzerland.

De Schutter, O. 2011b. "How Not to Think of Land-Grabbing: Three Critiques of Large-Scale Investments in Farmland." *Journal of Peasant Studies*, 38(2): 249–279.

Dhurua, S., and Gujar, G. 2011. "Field-Evolved Resistance to Bt Toxin Cry1Ac in the Pink Bollworm, *Pectinophora gossypiella* (Saunders) (Lepidoptera: Gelechiidae), from India." *Pest Management Science*, 67(8): 898–903.

Dibden, J., Gibbs, D., and Cocklin, C. 2013. "Framing GM Crops as a Food Security Solution." *Journal of Rural Studies*, 29: 59–70.

DOH (Hawaii State Department of Health). 2014. *2013–14 State Wide Pesticide Sampling Pilot Project Water Quality Findings*. Honolulu.

DOH (Hawaii State Department of Health). 2016. "Polluted Runoff Control Program." Clean Water Branch. http://health.hawaii.gov/cwb/site-map/clean-water-branch-home-page/polluted-runoff-control-program.

Douglas, M., and Tooker, J. 2015. "Large-Scale Deployment of Seed Treatments Has Driven Rapid Increase in Use of Neonicotinoid Insecticides and Preemptive Pest Management in US Field Crops." *Environmental Science and Technology*, 49(8): 5088–5097.

Dow AgroSciences. 2011. Petition for Determination of Nonregulated Status for Herbicide Tolerant DAS-40278–9 Corn. https://www.aphis.usda.gov/brs/aphisdocs/09_23301p.pdf.

Doyle, J. 2004. *Trespass against Us: Dow Chemical and the Toxic Century*. Monroe, ME: Common Courage Press.

DuPont. 2016. "The Science behind Feeding the World." http://www.dupont.com/corporate-functions/our-approach/global-challenges/food/articles/feeding-the-world.html.

DuPuis, M., and Goodman, D. 2005. "Should We Go 'Home' to Eat? Toward a Reflexive Politics of Localism." *Journal of Rural Studies*, 21(3): 359–371.

DuPuis, M., Goodman, D., and Harrison, J. 2006. "Just Values or Just Value? Remaking the Local in Agro-Food Studies." *Research in Rural Sociology and Development*, 12: 241–268.

Eagle, N. 2014. "GMO Issue Spurs Candidates to Seek Election, but It's Ugly Out There." *Civil Beat*. 5 August. http://www.civilbeat.com/2014/08/gmo-issue-spurs-candidates-to-seek-election-but-its-ugly-out-there-draft.

Eichenwald, K., Kolata, G., and Petersen, M. 2001. "Biotechnology Food: From the Lab to a Debacle." *New York Times*. 25 January. http://www.nytimes.com/2001/01/25/business/25FOOD.html.

Enay, S. 2011. "Hawaii's Tech Industry after Act 221." *Hawaii Business*. http://www.hawaiibusiness.com/hawaiis-tech-industry-after-act-221.

Engel, S., Wetmur, J., Chen, J., Zhu, C., Barr, D., Canefield, R., and Wolff, M. 2011. "Prenatal Exposure to Organophosphates, Paraoxonase 1, and Cognitive Development in Childhood." *Environmental Health Perspectives*, 119(8): 1182–1188.

Entine, J. 2013. "Hawaii GMO Investigation: Follow the Anti-Crop Biotech Money Trail." *Forbes*. 4 September. http://www.forbes.com/sites/jonentine/2013/09/04/hawaii-gmo-investigation-follow-the-anti-crop-biotech-money-trail/#613b08f23d44.

Eskenazi, B., Marks, A., Bradman, A., Harley, K., Bart, D., Johnson, C., . . . , and Jewell, N. 2007. "Organophosphate Pesticide Exposure and Neurodevelopment in Young Mexican-American Children." *Environmental Health Perspectives*, 115(5): 792–798.

ETC Group. 2013. "Putting the Cartel Before the Horse . . . and Farm, Seeds, Soil, Peasants, Etc." Communique No. 111.

Evenhuis, N., and Miller, S. (eds). 2015. *Records of the Hawai'i Biological Survey for 2014, Part II: Index*.

Feller, I. 1990. "Universities as Engines of R&D-Based Growth: They Think They Can." *Research Policy*, 19: 335–348.

Fernandez-Cornejo, J., and Schimmelpfennig, D. 2004. "Have Seed Industry Changes Affected Research Effort?" *Amber Waves*, 2(1): 14–19. Economic Research Service, USDA.

Finnerty, R. 2020. "Facing Economic Devastation, Hawaii Attempts to Revive Tourism." *All Things Considered*. National Public Radio. 20 October. https://www.npr.org/2020/10/20/925795410/facing-economic-devastation-hawaii-attempts-to-revive-tourism.

"Firmness, Once More." 1897. *Hawaiian Star.* 19 March. Accessed through *Chronicling America: Historic American Newspapers.* Library of Congress.

"Fish and Poi, Poi and Fish." 1867. *Pacific Commercial Advertiser.* 26 October. Accessed through *Chronicling America: Historic American Newspapers.* Library of Congress.

Fisher, M. 2009. *Capitalist Realism: Is There No Alternative?* Hants, UK: Zero Books.

Fisher, M., and Gilbert, J. 2013. "Capitalist Realism and Neoliberal Hegemony: A Dialogue." *New Formations: A Journal of Culture/Theory/Politics,* 80(80): 89–101.

Flynn, R. 2014. "Progress and Progressives in Hawai'i." *Farmers and Friends.* http://farmersandfriends.org/cool-free-reads/endangered-compact-progress-and-progressives-in-hawaii.

Forman, D. n.d. "Marooned in the Doldrums while Ignoring Indigenous Environmental Knowledge: Federal and State Preemption of County Ordinances Regulating Pesticide Use by GE/GMO Seed Companies in Hawai'i." Draft paper.

Foster, J., McChesney, R., and Jonna, R. 2011. "Monopoly and Competition in Twenty-First Century Capitalism." *Monthly Review,* 62(11): 1–39.

Freese, B., Lukens, A., and Anjomshoaa, A. 2015. *Pesticides in Paradise: Hawai'i's Health and Environment at Risk.* Hawai'i Center for Food Safety. www.centerforfoodsafety.org/reports.

Fridell, G. 2007. *Fair Trade Coffee: The Prospects and Pitfalls of Market-Driven Social Justice.* Toronto: University of Toronto Press.

Fujikane, C., and Okamura, J. (eds.). 2008. *Asian Settler Colonialism: From Local Governance to the Habits of Everyday Life in Hawai'i.* Honolulu: University of Hawai'i Press.

FWW (Food and Water Watch). 2012. *Public Research Private Gain.* Washington, DC.

FWW (Food and Water Watch). 2013. *Biotech Ambassadors: How the US State Department Promotes the Seed Industry's Global Agenda.* Washington, DC.

Garcia-Sancho, M. 2012. *Biology, Computing, and the History of Molecular Sequencing: From Proteins to DNA, 1945–2000.* New York: Palgrave Macmillan.

Gassmann, A., Petzold-Maxwell, J., Keweshan, R., and Dunbar, M. 2011. "Field-Evolved Resistance to Bt Maize by Western Corn Rootworm." *PloS one,* 6(7): 1–7.

Gibson, D. 2014. "Remembering the 'Big Five': Hawai'i's Constitutional Obligation to Regulate the Genetic Engineering Industry." *Asian-Pacific Law and Policy Journal,* 15(2): 213–283.

Gilbert, J. 2013. "What Kind of Thing Is 'Neoliberalism'?" *New Formations,* 80(80): 7–22.

Gillam, C. 2021. "Revealed: Monsanto Owner and US Officials Pressured Mexico to Drop Glyphosate Ban." *The Guardian.* 16 February. https://www.theguardian.com/business/2021/feb/16/revealed-monsanto-mexico-us-glyphosate-ban.

Gindin, S., Albo, G., Panitch, L., and Lilley, S. 2011. "Capitalist Crisis and Radical Renewal." In Lilley, S. (Ed.) *Capital and Its Discontents: Conversations with Radical Thinkers in a Time of Tumult.* Oakland, CA: PM Press. 105–122.

Giroux, H. 2014. *The Violence of Organized Forgetting: Thinking beyond America's Disimagination Machine.* San Francisco: City Lights Books.

Global Agriculture. 2017. "European Patent Office Continues to Grant Patents on Plants." Agriculture at a Crossroads. 4 July. https://www.globalagriculture.org/whats-new/news/en/32525.html.

Glover, D. 2010. "The Corporate Shaping of GM Crops as a Technology for the Poor." *Journal of Peasant Studies,* 37(1): 67–90.

Goldberger, J., Foltz, J., Barham, B., and Goeschl, T. 2005. "Modern Agricultural Science in Transition: A Survey of US Land Grant Agricultural and Life Scientists." University of Wisconsin, PATS Research Report 14.

Good Jobs First. n.d. "AccountableUSA—Hawaii." http://www.goodjobsfirst.org/states/hawaii.

Goodson, W., Lowe, L., Carpenter, D., Gilbertson, M., Ali, A., de Cerain Salsamendi, A., . . . , and Charles, A. 2015. "Assessing the Carcinogenic Potential of Low-Dose Exposures to Chemical Mixtures in the Environment: The Challenge Ahead." *Carcinogenesis*, 36(1), S254–S296.

Goodyear-Kaʻōpua, N., Hussey, I., and Wright, E. (eds.). 2014. *A Nation Rising: Hawaiian Movements for Life, Land, and Sovereignty*. Durham, NC: Duke University Press.

Gordines, J. 2013. "Bill 2491 Would Undo Great Work." *Garden Island*. 30 July. http://thegardenisland.com/news/opinion/guest/bill-would-undo-great-work/article_a4171b18-f8de-11e2-a9dc-0019bb2963f4.html.

Gorz, A. 2010. *The Immaterial: Knowledge, Value and Capital*. Translated by Chris Turner. London: Seagull Books.

Graeber, D. 2009. *Direct Action: An Ethnography*. Oakland, CA: AK Press.

Graeber, D. 2011. *Revolutions in Reverse*. New York: Minor Compositions.

Graeber, D. 2012. "Revolution at the Level of Common Sense." In Campagna, F., and Campiglio, E. (eds.). *What Are We Fighting For? A Radical Collective Manifesto*. London: Pluto Press. 165–175.

Graff, G., Rausser, G., and Small, A. 2003. "Agricultural Biotechnology's Complementary Intellectual Assets." *Review of Economics and Statistics*, 85(2): 349–363.

Gregg, A. 2008. "Teachers Drop TRO against Syngenta." *Garden Island*. 8 November. http://thegardenisland.com/news/teachers-drop-tro-against-syngenta/article_8b650731-8ee1-5ea5-8aed-1392650a33f3.html.

Grube, N. 2014. "Will the GMO Debate Fuel Campaign Donations as Local Elections Heat Up?" *Civil Beat*. 28 April. http://www.civilbeat.com/2014/04/21924-will-the-gmo-debate-fuel-campaign-donations-as-local-elections-heat-up.

Gupta, C. 2015. "Return to Freedom: Anti-GMO Aloha ʻĀina Activism on Molokai as an Expression of Place-Based Food Sovereignty." *Globalizations*, 12(4): 529–544.

Gurian-Sherman, D., and Mellon, M. 2014. *The Rise of Superweeds—and What to Do about It*. Union of Concerned Scientists, Policy Brief.

Guthman, J. 2004. *Agrarian Dreams? The Paradox of Organic Farming in California*. Berkeley: University of California Press.

Guthman, J. 2008. "Neoliberalism and the Making of Food Politics in California." *Geoforum*, 39(3): 1171–1183.

Hager, N. 2014. *Dirty Politics: How Attack Politics Is Poisoning New Zealand's Political Environment*. Nelson, NZ: Craig Potton Publishing.

Haiven, M., and Khasnabish, A. 2014. *The Radical Imagination: Social Movement Research in the Age of Austerity*. London: Zed Books.

Hanman, N. 2019. "Naomi Klein: 'We Are Seeing the Beginnings of the Era of Climate Barbarism.'" *The Guardian*. 14 September. https://www.theguardian.com/books/2019/sep/14/naomi-klein-we-are-seeing-the-beginnings-of-the-era-of-climate-barbarism.

Hansen-Kuhn, K. 2016. "Following Breadcrumbs: TPP Text Provides Clues to U.S. Positions in TTIP." Institute of Agriculture and Trade Policy, Minnesota. http://

www.iatp.org/documents/following-breadcrumbs-tpp-text-provides-clues-to-us-positions-in-ttip.
Hardt, M. 2011. "Reclaim the Common in Communism." *The Guardian*. 3 February. http://www.theguardian.com/commentisfree/2011/feb/03/communism-capitalism-socialism-property.
Hardt, M., and Negri, A. 2009. *Commonwealth*. Cambridge, MA: Harvard University Press.
Harmon, A. 2014. "A Lonely Quest for Facts on Genetically Modified Crops." *New York Times*. 4 January. http://www.nytimes.com/2014/01/05/us/on-hawaii-a-lonely-quest-for-facts-about-gmos.html.
Harvey, D. 1996. *Justice, Nature and the Geography of Difference*. Cambridge: Blackwell.
Harvey, D. 2005. *A Brief History of Neoliberalism*. Oxford: Oxford University Press.
Hasager, U., and Kelly, M. 2001. "Public Policy of Land and Homesteading in Hawai'i." *Social Process in Hawaii*, 40: 190–221.
Hauter, W. 2012. *Foodopoly: The Battle over the Future of Food and Farming in America*. New York: New Press.
Hayek, F. 1949. "The Intellectuals and Socialism." *University of Chicago Law Review*, 16(3): 417–433.
HDOA (Hawaii Department of Agriculture). 2015. "List of Active Special Local Need Registrations." Posted 30 January. http://hdoa.hawaii.gov/pi/files/2013/01/List-of-Active-SLNs-By-SLN-Number-with-Labels_01302015.pdf.
HDOA (Hawaii Department of Agriculture). 2016. "Agribusiness Development Corporation." http://hdoa.hawaii.gov/chair/adc10.
Henkin, D., and Moriwake, I. 2013. Legal Petition to the Commission on Water Resource Management of the State of Hawaii. Earthjustice, Attorney for Petitioner: Pō'ai Wai Ola/West Kaua'i Watershed Alliance. Honolulu.
Henry, M., Beguin, M., Requier, F., Rollin, O., Odoux, J., Aupinel, P., . . . , and Decourtye, A. 2012. "A Common Pesticide Decreases Foraging Success and Survival in Honey Bees." *Science*, 336(6079): 348–350.
Herbig, P., and Kramer, H. 1993. "High Tech Hawaii." *Journal of Business and Entrepreneurship*, 5(3): 107–119.
Herbig, P., and Kramer, H. 1994. "The Potential for High Tech in Hawai'i." *Social Process in Hawai'i*, 35: 58–70.
Hervey, T. 2012. "Boss GMO." *Honolulu Weekly*. 4 January. http://honoluluweekly.com/cover/2012/01/boss-gmo.
Hess, C. 1991. "Resource Allocation to State Agbiotech Research: 1982–1988." *Bio/Technology*, 9(1): 29–31.
Heynen, N., and Robbins, P. 2005. "The Neoliberalization of Nature: Governance, Privatization, Enclosure and Valuation." *Capitalism Nature Socialism*, 16(1): 5–8.
Higa, M. 2012. *Audit of the Department of Taxation's Administrative Oversight of High-Technology Business Investment and Research Activities Tax Credits*. Office of the Auditor, State of Hawaii, Report No. 12-05.
Hilbeck, A. 2008. "The IAASTD Report and Some of Its Fallout—a Personal Note." Institute of Integrative Biology, Zurich, Switzerland.
Hilbeck, A., Binimelis, R., Defarge, N., Steinbrecher, R., Székács, A., Wickson, F., . . . , and Wynne, B. 2015. "No Scientific Consensus on GMO Safety." *Environmental Sciences Europe*, 27(4): 1–6.

Hinrichs, C. 2003. "The Practice and Politics of Food System Localization." *Journal of Rural Studies*, 19(1): 33–45.
Hiraishi, K. 2018. "Hawai'i Supreme Court Hears Moloka'i Water Rights Case." Hawaii Public Radio. 31 May. https://www.hawaiipublicradio.org/post/hawai-i-supreme-court-hears-moloka-i-water-rights-case#stream/0.
Hofschneider, A. 2014a. "Federal Judge Invalidates Kauai's Anti-GMO Law." *Civil Beat*. 25 August. http://www.civilbeat.com/2014/08/court-declares-invalid-kauai-ordinance-regulating-gmos-pesticides.
Hofschneider, A. 2014b. "Lack of Money Leads to Lax Oversight of Pesticide Use in Hawaii." *Civil Beat*. 17 November. http://www.civilbeat.com/2014/11/lack-of-money-leads-to-lax-oversight-of-pesticide-use-in-hawaii.
Hofschneider, A. 2015a. "After 2 Years, Hawaii Still Won't Enforce Pesticide Disclosure Law." *Civil Beat*. 23 March. http://www.civilbeat.com/2015/03/after-2-years-hawaii-still-wont-enforce-pesticide-disclosure-law.
Hofschneider, A. 2015b. "Hawaii Is Feeling the Seed Industry's Downturn." *Civil Beat*. 24 June. http://www.civilbeat.com/2015/06/hawaii-is-feeling-the-seed-industrys-downturn.
Hofschneider, A. 2015c. "Hawaii to Expand Voluntary Pesticide Reporting by Big Ag Companies." *Civil Beat*. 9 October. http://www.civilbeat.com/2015/10/hawaii-to-expand-voluntary-pesticide-reporting-by-big-ag-companies.
Hofschneider, A., and Grube, N. 2014. "Taking Root: A Rising Voice in Hawaii's GMO Politics." *Civil Beat*. 30 April. http://www.civilbeat.com/2014/04/21937-taking-root-a-rising-voice-in-hawaiis-gmo-politics.
Holt-Gimenez, E., and Patel, R. (eds.). 2012. *Food Rebellions: Crisis and the Hunger for Justice*. Oakland, CA: Food First Books.
Holt-Gimenez, E., and Shattuck, A. 2011. "Food Crises, Food Regimes and Food Movements: Rumblings of Reform or Tides of Transformation?" *Journal of Peasant Studies*, 38(1): 109–144.
Hooser, G. 2015a. "9 Most Frequent Misstatements by Hawaii Chemical Companies." *GaryHooser's Blog*. 15 February. https://garyhooser.wordpress.com/2015/02/15/9-most-frequent-misstatements-by-hawaii-chemical-companies.
Hooser, G. 2015b. "Why I March for Aloha 'Āina—Join Me Sunday August 9 on Oahu." *GaryHooser's Blog*. 8 August. https://garyhooser.wordpress.com/2015/08/08/why-i-march-for-aloha-aina-join-me-sunday-august-9-on-oahu.
Hooser, G. 2015c. "State Law on RUP Disclosure Implemented. Kind of but Not Really—Plus Kauai County Stats on Glyphosate." *GaryHooser's Blog*. 9 September. https://garyhooser.wordpress.com/2015/09/09/state-law-on-rup-disclosure-implemented-kind-of-but-not-really-plus-kauai-county-stats-on-glyphosate.
Hooser, G. 2015d. "Trick or Treat? The Good Neighbor Program—A Masquerade of Disclosure." *GaryHooser's Blog*. 9 October. https://garyhooser.wordpress.com/2015/10/09/trick-or-treat-the-good-neighbor-program-a-masquerade-of-disclosure.
Hooser, G. 2017. "Paternalism, Power, and Politics." *GaryHooser's Blog*. 4 September. https://garyhooser.blog/2017/09/04/paternalism-power-and-politics-hooser-policy-politics.
Hooser, G. 2018. "Heroes of SB3095." *GaryHooser's Blog*. 22 June. https://garyhooser.blog/2018/06/22/heroes-of-sb3095-this-is-what-democracy-looks-like.
Horne, G. 2011. *Fighting in Paradise: Labor Unions, Racism, and Communists in the Making of Modern Hawai'i*. Honolulu: University of Hawai'i Press.

Horwitz, R., Finn, J., Vargha, L. and Ceaser, J. 1969. *Public Land Policy in Hawaii: An Historical Analysis*. Legislative Reference Bureau. Report No. 5. Honolulu: University of Hawaii.

Howard, P. 2018. "Global Seed Industry Changes Since 2013." 31 December. https://philhoward.net/2018/12/31/global-seed-industry-changes-since-2013.

Hubbard, K. 2009. *Out of Hand: Farmers Face the Consequences of a Consolidated Seed Industry*. Washington, DC: National Family Farm Coalition.

Hubbard, K. 2019. "The Sobering Details behind the Latest Seed Monopoly Chart." 11 January. https://civileats.com/2019/01/11/the-sobering-details-behind-the-latest-seed-monopoly-chart.

IAASTD. 2009. *Agriculture at a Crossroads*. International Assessment of Agricultural Knowledge, Science, and Technology for Development. http://www.unep.org/dewa/agassessment/reports/IAASTD/EN/Agriculture%20at%20a%20Crossroads_Global%20Report%20%28English%29.pdf.

International Survey of Herbicide Resistant Weeds. 2013. "Weeds Resistant to Glycines (G/9) by Species and Country." http://www.weedscience.org/summary/MOA.aspx?MOAID=12.

ISAAA Brief 55-2019. https://www.isaaa.org/resources/publications/briefs/55/executivesummary/default.asp.

Isaki, B. 2008. *A Decolonial Archive: The Historical Space of Asian Settler Politics in a Time of Hawaiian Nationhood*. Doctoral dissertation, University of Hawai'i.

ISB (Information Systems for Biotechnology). 2015. Searchable GE crop data. Sponsored by the USDA. http://www.nbiap.vt.edu.

Ishii-Eiteman, M. 2009. "Feeding the World, Greening the Planet." Pesticide Action Network North America, San Francisco.

Jaffee, D., and Howard, P. 2010. Corporate Cooptation of Organic and Fair Trade Standards. *Agriculture and Human Values*, 27(4): 387–399.

Jarves, J. 1838. "Sketches of Kauai." *The Hawaiian Spectator*, 1(1). Conducted by An Association of Gentlemen. Honolulu: Edwin O. Hall Printer. 66–86.

Jasanoff, S. 2005. *Designs on Nature*. Princeton, NJ: Princeton University Press.

Jasanoff, S. 2014. "Biotechnology and Empire: The Global Power of Seeds and Science." *Global Politics of Science and Technology*, 1(2): 201–225.

Jervis, G., and K. Smith. 2013. "Presentation by plaintiffs' attorneys in lawsuit by Waimea, Kaua'i residents against Pioneer, DuPont." 13 July. http://vimeo.com/70580803.

JFFG (Joint Fact Finding Group). 2016. *Pesticide Use by Large Agribusiness On Kaua'i*. Prepared by Accord 3.0 Network. http://www.accord3.com/pg1000.cfm.

JFFG (Joint Fact Finding Group) Draft. 2016. Draft Report and Recommendations. Released 10 May 2016.

Johnson, N. 2014. "Here's Why Hawaii's Anti-GMO Laws Matter" *Grist*. 20 November. http://grist.org/food/heres-why-hawaiis-anti-gmo-laws-matter.

Jung, M. 2010. *Reworking Race: The Making of Hawai'i's Interracial Labor Movement*. New York: Columbia University Press.

Kahane, J., and Mardfin, J. 1987. *The Sugar Industry in Hawaii: An Action Plan*. State of Hawaii Legislative Reference Bureau, Report No. 9. Honolulu.

Kamakau, S. 1992. *Ruling Chiefs of Hawaii*. Honolulu: Kamehameha Schools Press.

Kame'eleihiwa, L. 1992. *Native Land and Foreign Desires: Pehea Lā E Pono Ai?* Honolulu: Bishop Museum Press.

Kamiya, J. Ongoing. *HawaiiFarmersDaughter.* https://hawaiifarmersdaughter.com/.
Kanehe, L. 2014. "Kūʻē Mana Māhele: The Hawaiian Movement to Resist Biocolonialism." In Goodyear-Kaʻōpua, N., Hussey, I., and Wright, E. (eds.), *A Nation Rising: Hawaiian Movements for Life, Land, and Sovereignty.* Durham, NC: Duke University Press. 331–353.
Kaneya, R. 2015. "Hawaii Governor Declares State of Emergency for Homelessness." *Civil Beat.* 16 October. https://www.civilbeat.org/2015/10/hawaii-governor-declares-state-of-emergency-for-homelessness/.
Keim, B. 2012. "New GM Crops Could Make Superweeds Even Stronger." *Wired.* 1 May. http://www.wired.com/2012/05/new-superweed-evolution.
Kelly, M. 1994. "Foreign Investment in Hawaiʻi." *Social Process in Hawaii,* 35: 15–39.
Kenoi, B. 2013. "Mayor Kenoi Signs Bill 113." http://hawaiicountymayor.com/2013/12/05/mayor-kenoi-signs-bill-113.
Kent, N. 1993. *Hawaiʻi: Islands under the Influence.* Honolulu: University of Hawaiʻi Press.
Kent, N. 1994. "The End of the American Age of Abundance: Whither Hawaiʻi?" *Social Process in Hawaii,* 35: 179–194.
Kerr, K. 2014. "Pro-GMO Companies Spend $8 Million to Fight Maui Initiative." *Hawaii News Now.* 28 October. http://www.hawaiinewsnow.com/story/27106705/pro-gmo-companies-spend-8-million-to-fight-maui-initiative.
Kester, K. 2014. "Nixing Kauaʻi's Anti-GMO Ordinance Was Good for Hawaiʻi." *Civil Beat.* 2 September. http://www.civilbeat.com/2014/09/nixing-Kauaʻis-anti-gmo-ordinance-was-good-for-hawaii.
Khan, L. 2013. "How Monsanto Outfoxed the Obama Administration." *Salon.* 16 March. http://www.salon.com/2013/03/15/how_did_monsanto_outfox_the_obama_administration.
Khanna, A. 2012. "Seeing Citizen Action through an 'Unruly' Lens." *Development,* 55(2): 162–172.
Kiers, E., Leakey, R., Izac, A., Heinemann, J., Rosenthal, E., Nathan, D., and Jiggins, J. 2008. "Agriculture at a Crossroads." *Science,* 320(5874): 320–321.
Kimbrell, A. 1996. "Biocolonization: The Patenting of Life and the Global Market in Body Parts." In Mander, J., and Goldsmith, E. (eds.), *The Case against the Global Economy and for a Turn toward the Local.* San Francisco: Sierra Club Books.
Klein, N. 2014. *This Changes Everything: Capitalism vs. the Climate.* New York: Simon and Schuster.
Kloor, K. 2012. "GMO Opponents Are the Climate Skeptics of the Left." *Slate.* 26 September. http://www.slate.com/articles/health_and_science/science/2012/09/are_gmo_foods_safe_opponents_are_skewing_the_science_to_scare_people_.html.
Kloppenburg, J. 2004. *First the Seed: The Political Economy of Plant Biotechnology.* Madison: University of Wisconsin Press.
Kloppenburg, J. 2010. "Impeding Dispossession, Enabling Repossession: Biological Open Source and the Recovery of Seed Sovereignty." *Journal of Agrarian Change,* 10(3): 367–388.
Knutson, J. 2013. "Shortchanging Ag Research?" *AgWeek.* 21 January. http://www.agweek.com/crops/3790247-shortchanging-ag-research.
Koberstein, P. 2014. "GMO Companies Are Dousing Hawaiian Island with Toxic Pesticides." *Grist.* 16 June. http://grist.org/business-technology/gmo-companies-are-dousing-hawaiian-island-with-toxic-pesticides.

Koberstein, P., and Murphy, E. 2015. "The Silencing of Hector Valenzuela." *Cascadia Times*. 19 May. http://times.org/2015/05/19/the-silencing-of-hector-valenzuela.
Krimsky, S. 2005. "From Asilomar to Industrial Biotechnology: Risks, Reductionism and Regulation." *Science as Culture*, 14(4): 309–323.
Krimsky, S., Rothenberg, L., Stott, P., and Kyle, G. 1996. "Financial Interests of Authors in Scientific Journals: A Pilot Study of 14 Publications." *Science and Engineering Ethics*, 2(4): 395–410.
Kryder, D., Kowalski, S., and Krattiger, A. 2000. "The Intellectual and Technical Property Components of Pro-Vitamin A Rice (GoldenRice™): A Preliminary Freedom-to-Operate Review." University of New Hampshire School of Law. https://perma.cc/K3K9-4QQE.
"Kauai Good Neighbor Program Database." n.d. Hawaii Department of Agriculture. Accessed August 15, 2016. https://hdoa.hawaii.gov/pi/good-neighbor-data/. Site discontinued.
Landrigan, R., and Benbrook, C. 2015. "GMOs, Herbicides, and Public Health." *New England Journal of Medicine,* 373(8): 693–695.
Lave, R., Mirowski, P., and Randalls, S. 2010. "Introduction: STS and Neoliberal Science." *Social Studies of Science*, 40(5): 659–675.
Lawrence, G., Lyons, K., and Wallington, T. 2010. *Food Security, Nutrition, and Sustainability*. London: Earthscan.
Leacock, E. 1987. "Theory and Ethics in Applied Urban Anthropology." In Mullings, L. (Ed.) *Cities of the United States*. New York: Columbia University Press. 317–336.
Lee, W. 1850. Address of William Lee. Transactions of the Royal Hawaiian Agricultural Society, 1(1): 22–35.
Leong, C. 1997. " 'You Local or What?' An Exploration of Identity in Hawai'i." Doctoral dissertation, Graduate School of the Union Institute, Ann Arbor, Michigan.
Lerro, C., Hofmann, J., Andreotti, G., Koutros, S., Parks, C., Blair, A., . . . & Beane Freeman, L. 2020. "Dicamba Use and Cancer Incidence in the Agricultural Health Study: An Updated Analysis." *International Journal of Epidemiology*, 49(4): 1326–1337.
Leung, P., and Loke, M. 2008. "Economic Impacts of Increasing Hawai'i's Food Self-Sufficiency." University of Hawai'i College of Tropical Agriculture and Human Resources Cooperative Extension Service, Economic Issues paper, December.
Levy, H. 1966. *Industrial Germany: A Study of its Monopoly Organisations and Their Control by the State*. London: Routledge.
Li, Q., Wang, J., and Boesh, R. 2013. *Final Project Report for Kaua'i Air Sampling Study*. University of Hawai'i Department of Molecular Bioscience and Bioengineering.
Lilley, S. (ed.) 2011. *Capital and Its Discontents: Conversations with Radical Thinkers in a Time of Tumult*. Oakland, CA: PM Press.
Lincoln, N., and Vitousek. P. 2017. "Indigenous Polynesian Agriculture in Hawai'i." In *Oxford Research Encyclopedia of Environmental Science*. Oxford: Oxford University Press.
Lind, A. 1968. *An Island Community: Ecological Succession in Hawaii*. Westport, CT: Greenwood Press.
Lotter, D. 2008. "The Genetic Engineering of Food and the Failure of Science—Part 2: Academic Capitalism and the Loss of Scientific Integrity." *International Journal of Sociology of Agriculture and Food*, 16(1): 50–68.

Loudat, T., and Kasturi, P. 2009. *Hawaii's Seed Crop Industry: Current and Potential Economic and Fiscal Contributions.* Hawaii Crop Improvement Association and Hawaii Farm Bureau Federation.

Loudat, T., and Kasturi, P. 2013. *Hawai'i's Seed Crop Industry: Current and Potential Economic and Fiscal Contributions,* 2013 edition. http://www.hciaonline.com/hawaiis-seed-crop-industry-current-and-potential-economic-and-fiscal-contributions.

Lyte, B. 2015. "A Day at Syngenta." *Garden Island.* 13 October. http://thegardenisland.com/news/local/a-day-at-syngenta/article_8187854f-3f88-5520-a2eb-4c95e5be7dac.html.

MacLennan, C. 2014. *Sovereign Sugar: Industry and Environment in Hawai'i.* Honolulu: University of Hawai'i Press.

Magdoff, F., Foster, J., and Buttel, F. (eds.). 2000. *Hungry for Profit: The Agribusiness Threat to Farmers, Food, and the Environment.* New York: NYU Press.

Malkan, S. 2016. "Why Is Cornell University Hosting a GMO Propaganda Campaign?" *The Ecologist.* 22 January. http://www.theecologist.org/News/news_analysis/2986952/why_is_cornell_university_hosting_a_gmo_propaganda_campaign.html.

Manfredi, C. 2014. Hawaii Farm Bureau letter to members. 21 August.

Markowitz, G., and Rosner, D. 2003. *Deceit and Denial: The Deadly Politics of Industrial Pollution.* Berkeley: University of California Press.

Marx, K. 1992. *Capital,* vol. 1. London: Penguin Classics.

Mascarenhas, M., and Busch, L. 2006. "Seeds of Change: Intellectual Property Rights, Genetically Modified Soybeans, and Seed Saving in the United States." *Sociologia Ruralis,* 46(2): 122–138.

Matson, J., Tang, M., and Wynn, S. 2012. "Intellectual Property and Market Power in the Seed Industry: The Shifting Foundation of Our Food System." University of Wisconsin Law School, Government and Legislative Clinic.

Mazzucato, M. 2013. *The Entrepreneurial State: Debunking Public vs. Private Sector Myths.* London: Anthem Press.

McAfee, K. 2003. "Neoliberalism on the Molecular Scale: Economic and Genetic Reductionism in Biotechnology Battles." *Geoforum,* 34(2): 203–219.

McCarthy, J. 2004. "Privatizing Conditions of Production: Trade Agreements as Neoliberal Environmental Governance." *Geoforum,* 35(3): 327–341.

McCarthy, J., and Prudham, S. 2004. "Neoliberal Nature and the Nature of Neoliberalism." *Geoforum,* 35(3): 275–283.

McClain, D. 1992. "Hawaii's Competitiveness." In Roth, R. (Ed.) *The Price of Paradise,* Vol. I. Honolulu: Mutual Publishing. 7–14.

McGregor, D. 2002. "Kaho'olawe: Rebirth of the Sacred." *Amerasia Journal,* 28(3): 68–83.

McGregor, D. 2007. *Nā Kua'āina: Living Hawaiian Culture.* Honolulu: University of Hawai'i Press.

McMillan, G., Narin, F., and Deeds, D. 2000. "An Analysis of the Critical Role of Public Science in Innovation: The Case of Biotechnology." *Research Policy,* 29: 1–8.

Melrose, J., Perroy, R., and Cares, S. 2016. *Statewide Agricultural Land Use Baseline 2015.* Prepared for the Hawaii Department of Agriculture.

Meyer, M. 2014. "Hoea Ea: Land Education and Food Sovereignty in Hawaii." *Environmental Education Research,* 20(1): 98–101.

Mgbeoji, I. 2014. *Global Biopiracy: Patents, Plants and Indigenous Knowledge*. Vancouver: University of British Columbia Press.

Miles, A., and Merrigan, K. 2020. "If We Get Food Right, We Get Everything Right." *Civil Beat*, 11 April. https://www.civilbeat.org/2020/04/if-we-get-food-right-we-get-everything-right.

Mirowski, P. 2011. *Science-Mart*. Cambridge, MA: Harvard University Press.

Misalucha, B. 2015. "Pesticides Have Wide-Ranging Benefits." *Star-Advertiser*. 22 February.

Mitra, N. 2014. "Feeding Hawaii." *Earth Island Journal*. 18 June. http://www.earthisland.org/journal/index.php/elist/eListRead/feeding_hawaii.

Molden, D. (ed.) 2007. *Water for Food, Water for Life: A Comprehensive Assessment of Water Management in Agriculture*. London: Earthscan.

Monbiot, G. 2017. *Out of the Wreckage: A New Politics for an Age of Crisis*. London: Verso.

"Monsanto & Dow Spend More than $8 Million." 2014. Shaka Movement. 9 November. https://www.mauigmomoratoriumnews.org/project/unicorns-say-no.

"Monsanto to Plead Guilty to Criminal Count of Spraying Banned Pesticide." 2019. *Hawaii News Now*. 21 November. https://www.hawaiinewsnow.com/2019/11/22/monsanto-plead-guilty-criminal-count-spraying-banned-pesticide-maui.

Montgomery, E. 2012. "Genetically Modified Plants and Regulatory Loopholes and Weaknesses under the Plant Protection Act." *Vermont Law Review*, 37: 351–369.

Mooney, P. 2015. "The Changing Agribusiness Climate: Corporate Concentration, Agricultural Inputs, Innovation, and Climate Change." *Canadian Food Studies/La Revue canadienne des études sur l'alimentation*, 2(2): 117–125.

Moriwake, I. 2017. "Waimea Water Deal Is a 'Win-Win-Win' for Hawai'i." *EarthJustice Blog*, 5 May. https://earthjustice.org/blog/2017-may/waimea-water-deal-is-a-win-win-win-for-hawai-i.

Mortensen, D., Egan, F., Maxwell, B., Ryan, M., and Smith, R. 2012. "Navigating a Critical Juncture for Sustainable Weed Management." *BioScience*, 62(1): 75–84.

Moss, D. 2009. *Transgenic Seed Platforms: Competition Between a Rock and a Hard Place?* American Antitrust Institute. 23 October.

Moss, D. 2013. "Competition, Intellectual Property Rights, and Transgenic Seed." *South Dakota Law Review*, 58: 543–559.

Mugwanya, N. 2019. "After Agroecology: Why Traditional Agricultural Practices Can't Transform African Agriculture." *Breakthrough Institute*, 4 February 2019. https://thebreakthrough.org/journal/no-10-winter-2019/after-agroecology.

Mulvey, K., and Shulman, S. 2015. *The Climate Deception Dossiers: Internal Fossil Fuel Industry Memos Reveal Decades of Corporate Disinformation*. Union of Concerned Scientists.

Nagaoka, S. 2014. "Kauai's Good Neighbor Program Expresses Spirit of Collaboration." *Star-Advertiser*. 10 September. http://www.staradvertiser.com/editorialspremium/20140910--Kauais-Good-Neighbor-program-expresses-spirit-of-collaboration?id=274577771.

Narayan, S., and Scandrett, E. 2014. "Science in Community Environmental Struggles: Lessons from Community Environmental Monitors, Cuddalore, Tamil Nadu." *Community Development Journal*, 49(4): 557–572.

Newell, P. 2008. "Lost in Translation? Domesticating Global Policy on Genetically Modified Organisms: Comparing India and China." *Global Society*, 22(1): 115–136.

Newell, P. 2009. "Bio-Hegemony: The Political Economy of Agricultural Biotechnology in Argentina." *Journal of Latin American Studies*, 41(1): 27–57.

Newell, P., and Glover, D. 2003. "Business and Biotechnology: Regulation and the Politics of Influence." In Jansen, K., and Vellema, S. (eds.), *Agribusiness and Society: Corporate Responses to Environmentalism, Market Opportunities and Public Regulation*. London: Zed Books. 200–231.

Newfield, C. 2008. *Unmaking the Public University: The Forty-Year Assault on the Middle Class*. Cambridge, MA: Harvard University Press.

Okamura, J. 2008. *Ethnicity and Inequality in Hawai'i*. Philadelphia: Temple University Press.

Oki, D. 2003. *Surface Water in Hawaii*. US Department of the Interior, US Geological Survey.

Oreskes, N., and E. Conway, E. 2010. *Merchants of Doubt: How a Handful of Scientists Obscured the Truth on Issues from Tobacco Smoke to Global Warming*. New York: Bloomsbury Publishing.

Osorio, J. 2002. *Dismembering Lāhui: A History of the Hawaiian Nation to 1887*. Honolulu: University of Hawai'i Press.

Oxfam. 2019. "Public Good or Private Wealth?" https://www.oxfam.org/en/research/public-good-or-private-wealth.

Pala, C. 2015. "Pesticides in Paradise: Hawaii's Spike in Birth Defects Puts Focus on GM Crops." *The Guardian*, 23 August. http://www.theguardian.com/us-news/2015/aug/23/hawaii-birth-defects-pesticides-gmo.

Panitch, L., and Gindin, S. 2013. *The Making of Global Capitalism*. London: Verso Books.

Panitch, L., Henwood, D., and Lilley, S. 2011. "Demystifying Globalization." In Lilley, S. (Ed.) *Capital and Its Discontents: Conversations with Radical Thinkers in a Time of Tumult*. Oakland, CA: PM Press. 79–85.

Pape, E. 2015a. "Living Hawaii: The American Dream Is Broken for the Middle Class." *Civil Beat*. 11 February. http://www.civilbeat.com/2015/02/living-hawaii-the-american-dream-is-broken-for-the-middle-class.

Pape, E. 2015b. "Living Hawaii: The Nation's Worst Place to Earn a Living." *Civil Beat*. 6 July. http://www.civilbeat.com/2015/07/living-hawaii-the-nations-worst-place-to-earn-a-living.

Pape, E. 2015c. "Living Hawaii: Survey Finds Honolulu One of the World's Costliest Cities." *Civil Beat*. 22 September. http://www.civilbeat.com/2015/09/living-hawaii-survey-finds-honolulu-one-of-the-worlds-costliest-cities.

Parfitt, C. 2013. "How Are Genetic Enclosures Shaping the Future of the Agrifood Sector?" *New Zealand Sociology*, 28(4): 33–58.

Patel, R. 2013. "The Long Green Revolution." *Journal of Peasant Studies*, 40(1): 1–63.

Patel, R., and Goodman, J. 2019. "A Green New Deal for Agriculture." *Jacobin*. 4 April. https://jacobinmag.com/2019/04/green-new-deal-agriculture-farm-workers.

Paul, H., and Steinbrecher, R. 2003. *Hungry Corporations: Transnational Biotech Companies Colonise the Food Chain*. London: Zed Books.

Peck, J., and Tickell, A. 2002. "Neoliberalizing Space." *Antipode*, 34(3): 380–404.

Perry, T. 2013. Oral testimony submitted to Kauai County Council for Bill 2491. 9 October.

Petranek, L. 2001. "Will the Task Masters of the New Economy Please Stand Up!" *Social Process in Hawaii*, 40: 1–35.

Phillipson, M. 2015. Written testimony for Syngenta Hawaii, House Bill 1514. 19 February.

Philpott, T. 2018a. "This Weed Killer Is Wreaking Havoc on America's Crops." *Mother Jones*. https://www.motherjones.com/environment/2018/01/dicamba-monsanto-herbicide-neighbor-farms-soybeans.

Philpott, T. 2018b. "Bayer Bought Monsanto and Is Now Stuck with Its Biggest Headache." *Mother Jones*. https://www.motherjones.com/food/2018/06/bayer-bought-monsanto-and-is-now-stuck-with-its-biggest-headache.

Plasch, B. 1981. *Hawaii's Sugar Industry: Problems, Outlook, and Urban Growth Issues*. Hawaii State Department of Planning and Economic Development, Honolulu.

Pollack, A. 2009. "Crop Scientists Say Biotechnology Seed Companies Are Thwarting Research." *New York Times*. 19 February. http://www.nytimes.com/2009/02/20/business/20crop.html.

Pollack, A. 2013. "Unease in Hawai'i's Corn Fields." *New York Times*. 7 October. http://www.nytimes.com/2013/10/08/business/fight-over-genetically-altered-crops-flares-in-hawaii.html.

Pretty, J., Noble, A., Bossio, D., Dixon, J., Hine, R., Penning de Vries, F., and Morison, J. 2006. "Resource-Conserving Agriculture Increases Yields in Developing Countries." *Environmental Science and Technology*, 40(4): 1114–1119.

Preza, D. 2010. "*The Emperical Writes Back: Re-Examining Hawaiian Dispossession Resulting from the Māhele of 1848*." MA thesis, University of Hawai'i at Manoa.

Public Testimony RE: Kauai County Council Bill 2491 RELATING TO: Pesticides and Genetically Modified Organisms.

Public Testimony RE: Senate Bill No. 3170 RELATING TO: Pesticide Spraying Safety Zone, Submitted 4 February 2008. Name left out for privacy.

Ramones, I. 2014. "The True Cost of Hawai'i's Militarization." *Hawai'i Independent*. 11 September 2014. http://hawaiiindependent.net/story/the-true-cost-of-hawaiis-militarization.

Rancière, J. 2006. *Hatred of Democracy*. Translated by Steve Corcoran. London: Verso.

Rancière, J. 2010. *Dissensus: On Politics and Aesthetics*. London: Continuum.

Redfeather, N. 2012. "GMOs in Hawai'i—The Big Picture." In Black, C. (Ed.) *Facing Hawaii's Future*. Koloa, HI: Hawai'i Seed. 25–27.

Reed, G. 2018. "Peter Wright's Nomination Means Superfund Conflicts of Interest in Almost All 50 States." *Union of Concerned Scientists*. 16 April. https://blog.ucsusa.org/genna-reed/peter-wrights-nomination-superfund-conflicts-of-interest/.

Ritte, W. 2015. "Aloha 'Āina Stories from the Front Lines of Culture, Conservation and Community Activism." Presentation at Hawai'i Conservation Conference. Video by Occupy Hawaii. https://www.youtube.com/watch?v=YDZ-DcEuZro.

Roberts, J., Karr, C., Paulson, J., Brock-Utne, A., Brumberg, H., Campbell, C., . . . , and Wright, R. 2012. "Pesticide Exposure in Children." *Pediatrics*, 130(6): e1765–e1788.

Robinson, W. 2007. "The Pitfalls of Realist Analysis of Global Capitalism: A Critique of Ellen Meiksins Wood's Empire of Capital." *Historical Materialism*, 15(3): 71–93.

Rohrer, J. 2012. " 'Us vs Them': A Simple Recipe to Prevent Strong Society from Forming." *AlterNet*. 27 July. http://www.alternet.org/belief/us-vs-them-simple-recipe-prevent-strong-society-forming.

Rohter, I. 1992. *A Green Hawaii: Sourcebook for Development Alternatives*. Honolulu: Nā Kāne o ka Malo Press.

Rohter, I. 1994. "The Green Economy for Hawai'i." *Social Process in Hawaii*, 35: 24–144.

Rohter, I. 2001. "Fruits of Resistance: Hawaii Islanders Stop Oji Paper Ltd." Unpublished paper. University of Hawai'i.

Ross, E. 1998. *Malthus Factor.* London: Zed Books.
Roth, R. (Ed.) 1992. *The Price of Paradise,* Vol. I. Honolulu: Mutual Publishing.
Roth, R. (Ed.) 1993. *The Price of Paradise.* Vol. II. Honolulu: Mutual Publishing.
Sahagun, L. 1994. "Bitter End to Hawaii's 'Sugar Life.'" *Los Angeles Times.* 19 March. http://articles.latimes.com/1994-03-19/news/mn-35842_1_cane-fields.
Sai, K. 2008. The American Occupation of the Hawaiian Kingdom: Beginning the Transition from Occupied to Restored State. PhD thesis, University of Hawai'i at Mānoa.
Sass, J., and Wu, M. n.d. "US Pesticide Regulation: Weaknesses, Loopholes, and Flaws Undermine Farmworker Health." Environmental Health Policy Institute. http://www.psr.org/environment-and-health/environmental-health-policy-institute/responses/us-pesticide-regulation.html.
Sassen, S. 2010. "A Savage Sorting of Winners and Losers: Contemporary Versions of Primitive Accumulation." *Globalizations,* 7(1–2): 23–50.
Save Kauai Farms. 2013. YouTube, https://www.youtube.com/channel/UCUg4l4p0ur8zG9ydEQZuPaA?app=desktop.
Scheuer J., and Clark T. 2001. "Conserving Biodiversity in Hawai'i: What Is the Policy Problem?" *Yale F&ES Bulletin,* 105: 159–183.
Schimmelpfennig, D., Pray, C., and Brennan, M. 2004. "The Impact of Seed Industry Concentration on Innovation: A Study of U.S. Biotech Market Leaders." *Agricultural Economics,* 30(2): 157–167.
Schnurr, M. 2013. "Biotechnology and Bio-hegemony in Uganda: Unraveling the Social Relations Underpinning the Promotion of Genetically Modified Crops into New African Markets." *Journal of Peasant Studies,* 40(4): 639–658.
Schrager, B. 2014. *Seeds from Paradise: The Rise of Hawaii's Seed Corn Industry.* Master's thesis, University of Hawai'i.
Schrager, B., and Suryanata, K. 2018. "Seeds of Accumulation: Molecular Breeding and the Seed Corn Industry in Hawai'i." *Journal of Agrarian Change,* 18(2): 370–384.
Seager, J. 2003. "Rachel Carson Died of Breast Cancer: The Coming of Age of Feminist Environmentalism." *Signs,* 28(3): 945–972.
Sell, S. 2003. *Private Power, Public Law: The Globalization of Intellectual Property Rights.* New York: Cambridge University Press.
Sell, S. 2011. "TRIPS Was Never Enough: Vertical Forum Shifting, FTAS, ACTA, and TPP." *Journal of Intellectual Property Law,* 18: 447–478.
Shand, H. 2012. "The Big Six: A Profile of Corporate Power in Seeds, Agrochemicals and Biotech." *Heritage Farm Companion,* Summer 2012: 10–15.
Shaw, A. 2016a. "Gendering in Seed-Agrichemical Companies: Workforce Composition, Divisions of Labour and Corporate Policies." Doctoral dissertation draft chapter. London School of Economics.
Shaw, A. 2016b. "TechnoPlantations: Gender, Race, and Labour in Hawai'i's Seed Production Economies." Unpublished paper. London School of Economics.
Shelton J., Geraghty E., Tancredi D., Delwiche L., Schmidt R., Ritz B., Hansen R., and Hertz-Picciotto, I. 2014. "Neurodevelopmental Disorders and Prenatal Residential Proximity to Agricultural Pesticides: The CHARGE Study." *Environmental Health Perspectives,* 122(10): 1103–1110.
Siegel, L. 2011. "Brownfields in 'Paradise': Kekaha's Legacy of Industrial Agriculture." Center for Public Environmental Oversight. Mountain View, California.

Silva, N. 2004. *Aloha Betrayed: Native Hawaiian Resistance to American Colonialism.* Durham, NC: Duke University Press.

Skolnick, A. 2013. "GMOs Are Tearing a Tropical Paradise Apart." *Salon.* 5 September. http://www.salon.com/2013/09/04/a_battle_in_paradise_how_gmos_are_tearing_a_tropical_utopia_apart.

Smith, E., Azoulay, D., and Tuncak, B. 2015. *Lowest Common Denominator: How the Proposed EU-US Trade Deal Threatens to Lower Standards of Protection from Toxic Pesticides.* Center for International Environmental Law, Washington, DC.

Southern Poverty Law Center. 2013. *Close to Slavery: Guestworker Programs in the United States.* http://www.splcenter.org/get-informed/publications/close-to-slavery-guestworker-programs-in-the-united-states.

Sproat, K. 2009. *Ola I Ka Wai: A Legal Primer for Water Use and Management in Hawai'i.* Ka Huli Ao Center for Excellence in Native Hawaiian Law, Honolulu.

Sproat, D. 2011. "Where Justice Flows Like Water: The Moon Court's Role in Illuminating Hawai'i Water Law." *University of Hawai'i Law Review,* 33: 537–579.

Srinivasan, C. 2003. "Concentration in Ownership of Plant Variety Rights: Some Implications for Developing Countries. *Food Policy,*" 28(5): 519–546.

State of Hawaii Agribusiness Development Corporation Agreement No. L-08202, 2007. "LICENSE AGREEMENT NO. L-08202 between STATE OF HAWAII AGRIBUSINESS DEVELOPMENT CORPORATION as LICENSOR and SYNGENTA SEEDS, INC. a Delaware corporation as LICENSEE." Honolulu.

State of Hawaii Office of the Auditor. 2021. *Audit of the Agribusiness Development Corporation.* Report No. 20-01. Honolulu.

Stauffer, R. 2001. "The University of Hawai'i, Public Policy, and the Process of Globalization." *Social Process in Hawaii,* 40: 91–121.

Steingraber, S. 1997. *Living Downstream: An Ecologist Looks at Cancer and the Environment.* Boston: Addison-Wesley Publishing.

Steingraber, S. 2016. Comment letter on draft report, *Pesticide Use by Large Agribusiness on Kaua'i.* 8 April.

Stoltzfus, R. 2014. "GMO Warriors." *Farmers and Friends.* http://farmersandfriends.org/cool-free-reads/gmo-warriors.

Stop Poisoning Paradise. 2013. "Sample Letters from 20 Doctors and Nurses." https://www.stoppoisoningparadise.org/doctors-and-nurses-letters-to-mayor.

Swanson, D. 2014. "A New Estimation of the Hawaiian Population for 1778, the Year of First European Contact." Pp. 9–23 in *Ka Huaka'i: 2014 Native Hawaiian Educational Assessment.* Honolulu: Kamehameha Publishing.

Swyngedouw, E. 2007. "Impossible 'Sustainability' and the Postpolitical Condition." In Krueger, R., and Gibbs, D. (eds.), *The Sustainable Development Paradox: Urban Political Economy in the United States and Europe.* New York: Guilford Press. 13–40.

Swyngedouw, E. 2011. "Whose Environment? The End of Nature, Climate Change, and the Process of Post-Politicization." *Ambiente and Sociedade,* 14(2): 69–87.

"Technology Omnibus." 1999. Report to the State of Hawaii Senate. https://www.capitol.hawaii.gov/session1999/bills/SB1583_HD1_.htm.

Then, C., and Tippe, R. 2012. *European Patent Office at Crossroads.* Munich: No Patents on Seeds.

Trask, H. 1991. "Coalition-Building between Natives and Non-Natives." *Stanford Law Review,* 43(6): 1197–1213.

Trask, H. 1999. *From a Native Daughter: Colonialism and Sovereignty in Hawai'i.* Honolulu: University of Hawai'i Press.

Trask, M. 2006. "Hawaiian Perspectives on GMOs." In Currie, A. (Ed.) *Facing Hawai'i's Future*, 1st ed. Hawai'i: Mandala Publishing. 24–27.
Uechi, C. 2019. "Molokai Group Seeks to Restore Stream Flows." *Maui News*. 2 July. https://www.mauinews.com/news/local-news/2019/07/molokai-group-seeks-to-restore-stream-flows.
UNCTAD (United Nations Committee on Trade and Development). 2013. *Wake Up before It Is Too Late*. United Nations, Geneva, Switzerland.
USDA (United States Department of Agriculture). 2013. Dow AgroSciences Petitions for Determinations of Nonregulated Status for 2,4-D-Resistant Corn and Soybean Varieties, Draft Environmental Impact Statement.
USDA (United States Department of Agriculture). 2016. "Pacific Basin Agricultural Research Center—Research Programs and Projects at this Location." http://www.ars.usda.gov/research/projects_programs.htm?modecode=20-40-05-00.
USDA (United States Department of Agriculture). 2020. BRS Interstate/Release and Release Permits and Notifications. https://www.aphis.usda.gov/aphis/ourfocus/biotechnology/permits-notifications-petitions/sa_permits/status-update/release-permits.
USDA National Agricultural Statistics Service. 2020. "Pacific Region—Hawaii Seed Crops." Press release, 18 May. Honolulu.
USGAO (US Government Accountability Office). *Genetically Engineered Crops: Agencies Are Proposing Changes to Improve Oversight, but Could Take Additional Steps to Enhance Coordination and Monitoring*. Report to the Committee on Agriculture, Nutrition, and Forestry, US Senate.
Valenzuela, H. 2012. "*Environmental and Health Risks of Synthetic Chemicals Used by the Biotechnology Seed Industry in Hawaii*, Draft 2.0." Unpublished paper.
Valenzuela, H. 2016. Comments to the Kauai Joint Fact Finding Group. 7 April.
Vallianatos, E. 2014. *Poison Spring: The Secret History of Pollution and the EPA*. New York: Bloomsbury Publishing.
Van Dyke, J. 2008. *Who Owns the Crown Lands of Hawai'i?* Honolulu: University of Hawai'i Press.
Van Voorhis, V. 2011. "County Takes Legal Action against Seed Companies." *Garden Island*. 3 May. http://thegardenisland.com/news/local/county-takes-legal-action-against-seed-companies/article_efe173fa-762b-11e0-a0f1-001cc4c002e0.html.
Via Campesina. 1996. "The Right to Produce and Access to Land." Position of the Via Campesina on Food Sovereignty presented at the World Food Summit, Rome, Italy.
Virno, P. 2004. *A Grammar of the Multitude: For an Analysis of Contemporary Forms of Life*, Translated by Isabella Bertoletti, James Cascaito, and Andrea Casson. New York: Semiotext(e).
Vogeler, K. 2014. "Outside Shangri La: Colonization and the US Occupation of Hawai'i." In Goodyear-Ka'ōpua, N., Hussey I., and Kahunawaika'ala Wright, E. (eds.), *A Nation Rising: Hawaiian Movements for Life, Land, and Sovereignty*. Durham, NC: Duke University Press.
Voosen, P. 2011. "King Corn Takes Root in Hawaii." *New York Times*, 22 August. http://www.nytimes.com/gwire/2011/08/22/22greenwire-king-corn-takes-root-in-hawaii-28466.html.
Wayland, F. 1837. *The Elements of Political Economy*. Boston: Gould and Lincoln.
Whitehorn, P., O'Connor, S., Wackers, F., and Goulson, D. 2012. "Neonicotinoid Pesticide Reduces Bumble Bee Colony Growth and Queen Production." *Science*, 336(6079): 351–352.

Wise, H. 1850. *Los Gringos: An Inside View of Mexico and California, with Wanderings in Peru, Chili, and Polynesia.* New York: Baker and Scribner.
Witeck, J. 2001. "Public Policy in Hawaii: Globalism's Neoliberal Embrace." *Social Process in Hawaii,* 40: 36–68.
Wolin, S. 2008. *Democracy Incorporated: Managed Democracy and the Specter of Inverted Totalitarianism.* Princeton, NJ: Princeton University Press.
Wood, E. 2002. *The Origin of Capitalism: A Longer View.* London: Verso.
Wood, E. 2003. *Empire of Capital.* London: Verso.
Wood, E. 2012. *Liberty and Property. A Social History of Western Political Thought from the Renaissance to the Enlightenment.* London: Verso.
Wood, E., and Lilley, S. 2011. "Empire in the Age of Capital." In Lilley, S. (Ed.) *Capital and Its Discontents: Conversations with Radical Thinkers in a Time of Tumult.* Oakland: PM Press. 27–42.
Wynne, B. 1992. "Misunderstood Misunderstanding: Social Identities and Public Uptake of Science." *Public Understanding of Science,* 1(3): 281–304.
Wynne, B. 2005. "Reflexing Complexity: Post-Genomic Knowledge and Reductionist Returns in Public Science." *Theory, Culture, and Society,* 22(5): 67–94.
Xia, Y., and Buccola, S. 2005. "University Life Science Programs and Agricultural Biotechnology." *American Journal of Agricultural Economics,* 87(1): 287–293.
Yap, B. 2013. "Against the Grain?" *Mana Magazine,* July–August. http://www.welivemana.com/articles/against-grain.
Yerton, S. 2018. "Surprise Eviction Notices Have This Old Plantation Community up in Arms." *Civil Beat,* 19 October. https://www.civilbeat.org/2018/10/surprise-eviction-notices-have-this-old-plantation-community-up-in-arms.
Yerton, S. 2019. "Seed Industry Is Shrinking Dramatically but It Still Grows Hawai'i's No. 1 Crop." *Civil Beat,* 22 November. https://www.civilbeat.org/2019/11/seed-industrys-footprint-is-shrinking-dramatically-but-it-remains-hawaiis-no-1-crop.
Zerbe, N. 2015. "Plant Genetic Resources in an Age of Global Capitalism." *Canadian Food Studies/La Revue canadienne des études sur l'alimentation,* 2(2): 194–200.
Zinn, H. 1980. *A People's History of the United States.* New York: Harper Collins.
Zinn, H. 1990. *Declarations of Independence: Cross-Examining American Ideology.* New York: Perennial.
Zinn, H. 2002. *You Can't Be Neutral on a Moving Train: A Personal History of Our Times.* Boston: Beacon Press.
Zinn, H. 2012. *Howard Zinn Speaks: Collected Speeches 1963 to 2009.* Edited by Anthony Arnove. Chicago: Haymarket Books.

Index